Reading and working with John Samuel's framework for understanding my innate gifts and strengths has changed my life. Three years ago I was mired in a career direction that left me unengaged and dispirited at the end of each day. John's work helped me to discover and understand my gifts and innate strengths and to re-direct my career into a direction that gives expression to those gifts and is much more energizing. Never before have I understood myself and my purpose so completely. I recommend this to anyone who feels lost in their life direction.

– **Don Foster Ph.D**
Senior Research Scientist/Geneticist, Seattle, Washington, USA

Finding our life mission is the foundational cornerstone in the quest toward a meaningful life. Judaism teaches that in order to define our purpose in life we must distinguish who we are from what we do. John Samuel helped me recognize where and how to focus my energies in fulfilling my purpose as a spiritual leader. I believe this work will likewise help you define and implement your own "mission statement" and achieve the satisfaction of knowing that we make a positive contribution to our community and society.

– **Rabbi Mordechai Abergel**
Chief Rabbi, Maghain Aboth Synagogue, Singapore

Over the years I've taken numerous diagnostics, inventories, and personality profiles. While each has shed light on various aspects of my design, none has come close to the depth and spot-on accuracy that this research has uncovered. It has been both a revelation and a liberation for my soul. This book has given me the clarity to make decisions about how I can be most productive and effective. It is truly life-changing. I recommend it to all.

– **Stan Lander**
Broadcast Consultant, Seattle, Washington, USA

Over the past 5 years I have been able to reorient my business to the areas and manner in which the Talent analysis showed we would have the most success. It is therefore no coincidence that over this period we have emerged as one of the top 3 companies in our field in the country. Within the company too, we have realigned my responsibilities to the areas where I am achieving much more with much less effort. I salute John's insightful research! "

— **Suresh Williams**
President/CEO, Surya Gears Private Limited, India

Do we all have innate talents? John Samuel and Skip Moen write convincingly on the subject, and going through the process of Talent analysis gave my husband and me many helpful insights into ourselves and each other — many 'A-ha moments' — as to why we function as we do. This has been helpful in our not-for-profit work and also in our own relationship. I recommend this book for any who is seeking greater understanding about their own or others' capabilities.

— **Christine Patterson**
United Kingdom

This book outlines John's research journey and his anguished cry to find the secret of what makes anyone tick and culminates in the "Aha" moment when the insight into an individual's innate talent design is gained. This knowledge is a necessity.

— **Dr Hannah Anandaraj**
Chairperson, World Vision, India

Are there some things that do not change that can serve as anchors for our lives? John Samuel and Skip Moen challenge us to discover our specific "talent architecture", our "zone" and to choose work that allows us to express that zone. I have benefited from their work and encourage others to do so.

— **Rev Dr Soo Inn Tan**
Director, Graceworks, Singapore

"The important thing about the Phenomenon of the Zones is that it happens. We can fight it, but at the end of the day, who we are will shine through. Therefore, the content of this book is critical to understand and internalize. It is not only life changing personally, but it allows us to better understand our impact on others. In that way, we can be intentional about not only maximizing our own development, but also the development and success of those around us. Being armed with this reality of our lives, helps us to be intentional about living life with passion and purpose. What a gift the insight about Zones brings to the world!"

– Aaron Couch
Lead Pastor, Real Life Ministries, Moscow, Idaho, USA

"Increased profitability through better financial management? Already been conquered. Then how about enhanced productivity by leveraging technology? Done lots of that too. But so have all of our competitors. So what's left... what's the next "great breakthrough" in global competitiveness? It's the "people"! Try tapping into each person's fundamental gifting. Their deepest construct – that releases their utmost productive power. John Samuels & Skip Moen are onto something BIG here – worthy of our time and deep attention."

– Michael Smith
Management Consultant, Seattle, Washington, USA

"As a marriage and family therapist, I have come to yearn for some absolutes concerning the struggle we all have in knowing ourselves and the potential within. John Samuel has cracked the human passion-code and uncovered a phenomenon that I have witnessed first-hand to result in a powerful freedom to know and be known. Open this book and critically savor the personal truths within!"

– Dr. Doug Wheeler
Marriage and Family Therapist, Woodinville, Washington, USA

Living In Your Zone

Where Work Becomes Delight

John B. Samuel
& Skip Moen Ph.D

This book is dedicated to you.

It was birthed out of a passion to help you find yourself by uncovering your talent design, to help you appreciate your specific strengths *and* your natural limitations.

In the words of Bernard Haldane "Because you are unique, there is something you are better at than anybody else". Discovering your unique competitive advantage is foundational for a fruitful and fulfilling vocational career – and other important life engagements, family and social life included.

If this book will help initiate your journey of uncovering your talent design, it has served its purpose.

I am grateful to hundreds of my professional clients worldwide, in every continent spanning diverse national, racial, cultural and linguistic backgrounds. I am deeply indebted to each of them as they shared with me their life stories - and gave me both data and perspective about how much passion borne out of talent is the key to personal fulfillment, exceptional performance and human advancement.

And to each of my client's, I want to say "thank you" – without you this book would not have been possible!

John B. Samuel
March 2011

Talent Research Foundation
15000 Thoroughbred Lane
Montverde, FL 34756
USA
www.talentresearch.org

Living In Your Zone – Where Work Becomes Delight
Copyright © 2011 by John B. Samuel and Skip Moen

All rights reserved. Written permission must be secured from the publisher to use or reproduce any part of this book, except for brief quotations in critical reviews or articles.

ISBN: 978-0578088136

Cover Design: Juliann Itter, USA
Typeset Design: Ruiping Tan & Lee Tian Yi, Singapore

Contents

Introduction: What's Wrong With Work?

Chapter 1	Three for the Money: Work Woes	13
Chapter 2	Stumbling on the Solution – How We Uncovered the Principles Governing Human Productivity	25
Chapter 3	Myth Busting – Cleaning Up the Trash	33
Chapter 4	The DNA of Optimal Productivity – The Question of a Lifetime	43
Chapter 5	The Productivity Genome – The Double Helix of Work	59
Chapter 6	The Zone Phenomenon – Uncovering the Laws that Govern Human Productivity	69
Chapter 7	Zone 1 – Driven to Uncover the Latent and Hidden	73
Chapter 8	Zone 2 – Masterful Executor	81

Chapter 9	Zone 3 – The Articulate Teacher 89
Chapter 10	Zone 4 – The Quintessential Encourager .. 97
Chapter 11	Zone 5 – The Resource Manager 107
Chapter 12	Zone 6 – Builders and Entrepreneurs 117
Chapter 13	Zone 7 – A Passion to Meet Needs 129
Chapter 14	Uncovering Your Productivity Compass – Its Implications and Applications 139
Chapter 15	Rainbow Hunting 143
Chapter 16	Chasing Rainbows 151
Chapter 17	Don't Try This At Home 161

Epilogue: Now What Do I Do? 167

INTRODUCTION

WHAT'S WRONG WITH WORK

"I just don't like my job. I do it because I have to. I need to pay the bills."

That's the bottom line for most people today. The Conference Board[1] tells us that 70% of Americans do not like what they do for income. The number of people who are paid well for doing what they love are few and far between. We envy them. We wish it would happen to us. But we don't have any idea how to bring about that happy marriage of economic advantage and passionate fulfillment. So, we save for vacation. We endure work in order to go do something we love when we are not working. It's a terrible rat-race of futility. The more we commit to this kind of life, the greater the downside. Stress, health issues, boredom, discouragement, addictive release and all kinds of dysfunctional behaviors crop up in our lives.

> *Work has been divorced from who I really am, and in this divorce, everyone suffers.*

The problem isn't work. It's the wrong normal! We are committed to a "normal" view of life that sees work as production rather than fulfillment. Work is a part of the cogs of industry – a function of the mechanism of commerce. So, work is something that has to be done, not something that I love to do. Work has been divorced from who I really am, and in this divorce, everyone suffers.

It's time to re-examine the fundamentals of work – how my work reflects who I am. It's time to see work as an expression of my identity, not an application of my skills. In this book, we will look at many individuals who struggled with the basic question, "What was I born to do?" Through their struggles and discoveries, we will find that the normal view is flawed – fundamentally flawed. The normal view is bankrupt because it does not stem from a deep understanding of what makes each one of us unique in the world. Built on the philosophy of replacement parts, work today crushes the life out of us. It's time to bring vitality, enthusiasm and vibrancy back into our work. And we start by throwing out the old normal.

It's time to see work as an expression of my identity, not an application of my skills.

[1] www.conference-board.org

Chapter 1

Three for the Money: Work Woes

"You can't do that!" Dan felt like he'd just taken a sucker punch to the stomach. The adrenaline sent his heart rate soaring. His blood pressure jumped twenty points. He was dizzy.

"Now, take it easy, Dan. It's not like you're getting fired. We just need to move you to a different role in the company."

"But I built this company – from scratch. It's always been about research. That's what we do."

"Yes, we know, and we're grateful for your contribution. Without your work, we'd never be able to take this next step. But, you know, we're in a great position here. We can switch over to manufacturing and make a lot more money with a lot less investment. We get rid of all the research expense and reap the profits. You'll be set for life. Your shares will be worth millions. What's wrong with that?"

Dan hardly heard Charlie's answer. He thought about his love for the lab, a lab he'd built one piece of equipment at a time. He thought about the amazing hours investigating the intricate structures of DNA and the exhilaration of each new discovery.

"There has to be another way, Charlie. I mean, I'm the Chief

Scientific Officer here. What does a Chief Scientific Officer do without research?"

"Well, that's a problem, isn't it? I mean, technically, there's really no more research to do. We plan to offer you a consulting position – senior, of course. You'll be our spokesman to the industry, get to travel, go to all the conferences. You'll be free to do what you want."

"But I want to do research."

"Look, I know you love this stuff, but there's just too much money at stake here." Charlie paused. "Why don't you consider retiring? You'll get a fat golden handshake. You and Kathy can go visit Europe. You always talked about that. You'll have the time and plenty of money."

"But this is what I want!"

"I don't know what to say, Dan. The Board made their choice. We're closing this part down and shifting to manufacturing. We're looking at a great new plant in Michigan. You know, they're offering us some terrific tax breaks to go there. Look Dan, it's pretty simple, really. It's about the money. We just can't spend any more on pure research. It's time to reap the harvest."

Dan slumped at his desk. All those years. For what? To get rich? He didn't care about getting rich. He cared about finding the secrets. But the rest of the executive team only cared about profits. As far as they were concerned, what he did didn't matter anymore.

Dan Howard had an outstanding resume. With many critical

scientific discoveries in drug therapy, he was a recognized expert in the field. The author of more than one hundred scientific articles, holder of more than forty patents, Dan led the way in pure research. He loved it. Now, after years of investment and countless hours in the lab, he was being pushed aside by a change in corporate objectives. It made him sick.

Dan's story isn't unusual. Talent isn't measured by financial return. Money doesn't drive everyone. We recognize other motivations in life when we see musicians who play for pennies because they love to make music, artists who paint in poverty, often unrecognized until after their deaths or athletes who forego lucrative salaries in order to win a championship ring. Personally, we know that money is often not a motivator. We want to do what we love to do. We want to feel we are being fulfilled in our work, not just paid to perform. Far too many of us go to work because we have to pay the bills, longing for the days when we can do what really brings us joy. It is a sad state of affairs. Work isn't supposed to rob us of our passion. Just like Dan, we want our effort to result in something significant, something that has meaning for who we are.

> *We want to feel we are being fulfilled in our work, not just paid to perform.*

It might be a new patent in genetic research or simply a better way to tie a fishing fly, but whatever it is, it has to reflect what we were born to be if it's going to bring joy on the job. Dan's story is just one of the thousands of stories about people who have been displaced by economic pressure, removed from what they do best because someone could make more money without them. Unlike most people, Dan clearly understood where his talent was most productive. His problem was not clarity of purpose. It was misalignment with corporate direction. Like most people, Dan's talent wasn't recognized as an important part of work. It was simply something to be used until it no longer met the needs of the company. The decision that governed this value was strictly business, nothing personal – except, of course, that it was intensely personal for Dan and for the millions of others like him.

Dan's problem was not clarity of purpose. It was misalignment with corporate direction.

Henry Woo's story ...

Henry Woo was born in a family that had risen from the ashes. When it was clear Hong Kong would be returned to communist China, his father made the decision to leave. They immigrated to Vancouver with nothing in their collective pockets but the will to survive. For twenty years, Henry's father built a business by working every minute of every day. Now the firm was a prestigious accounting practice, one of the best in the city. The revenues flowed in from one corporate client after another. Years of hard labor were paying great dividends for everyone – except Henry.

"It's our family business. I spent my entire working life preparing for this day. I'm giving it to you on a silver platter. All those years that I suffered and scraped by were just so I could turn it all over to you. How can you be so ungrateful?"

Henry played that conversation in his head every day as he rode to the office. He could hear his father's intensity and passion. Yes, it was all true. When his parents arrived in this place, they had nothing. Year after year they denied themselves in order to pour everything into the business. Now the company was powerful and lucrative. His father wanted to retire. Henry could see the pride in his father's eyes as he recounted the years of toil needed to reach this level of success. But all that success, all those expectations on him, created nothing but panic and distress. He wanted to be an explorer. He wanted to feel the energy racing through his body as he pieced together the seemingly disconnected parts of a puzzle. He wanted to devote himself to his real talent instead of hiding it under a mountain of corporate meetings and paperwork. He might not carry a prestigious title. He might not ride in the chauffeured car. He might not live in a million-dollar house. But he knew life would be joyful instead of this constant dread. Was he going to be imprisoned in a life he didn't choose just because the family expected it? No, that could never be! But how could he turn away from all his father had done for him? How could he reject such a gift? In his culture, such a thing was simply unconscionable. He faced an impossible dilemma. A life of luxurious drudgery or a life of joyful rejection. Every morning brought the same internal debate. He just didn't know how much longer he could keep pretending.

Henry's older brother escaped the family expectations by taking an acceptable route. Now a member of one of the world's prestigious consulting firms, he could offer the legitimate excuse, "I'm upholding the family name in business. Everyone knows how valuable I am." With the older brother out of the picture, Henry's father turned his attention to the younger son. Sending him to the best school for his accounting degree, Henry performed admirably. He returned to Vancouver with the zest and drive to make the practice his own.

At first, the novelty and challenge propelled him. But the

day came, sooner rather than later, when he realized that the mountain of invoices and the piles of receipts to be reconciled was nothing but a nightmare of routine, laborious detail. As long as he was with a client, developing the relationship, solving the strategic problems, investigating issues, his spirit soared. Returning to the office was the dreaded reality. He loved being a forensic detective, combing through the financial evidence in order to find where the money really went. But the massive routine of ordinary accounting, the heart of the business, became a mixture of alcohol, stomach antacids and restless sleep. Now he was sure it would never end. His father's dream became Henry's nightmare.

Most working adults share some portion of Henry's dilemma. They find themselves operating in the second category of work. The first category is compulsory labor. For centuries, work was often associated with slavery. Production was the result of the strong arm of the master, either under a sovereign king or under the slave-owner. There was no choice for the slave. He either complied with the demands of the ruler or he suffered the consequences.

The second category of work now dominates production. Compulsion has been replaced with obligation. This is labor in the free market. We decide where, how and when we will work but we are driven to make those choices by necessity, not by passion. Today the vast majority are slaves to their own desires. They do not work because work itself is joyful. They work because the results of their labor provide other things that make life possible and, hopefully, joyful. In other words, they work to live, not live to work. It is the conversion of their labor into monetary exchange that drives them.

They want what others provide and they are willing to spend their lives in the pursuit of those desires by making the money necessary to purchase them. But far too often, the price they pay is much higher than the amount they spend at the cash register. The real cost is calculated in stress, illness, psychological depression and discouragement, broken relationships and that feeling of being stuck in a job you don't like but have to do. This category of work is far from those moments of genuine fulfillment that come when we are doing what we were born to do. Work because of obligation is a long way from work that results from passionate engagement. Just like Henry, most of us do what we have to in order to get by. We might work at the job because of family pressure and expectations.

They do not work because work itself is joyful. They work because the results of their labor provide other things that make life possible and, hopefully, joyful.

We might labor because of economic necessity or lifestyle demands. But the bottom line is the same: we do a job we know we weren't born to do – and we hate every minute of it.

Peter Tan's story ...

"Why me?" That seemed to be the constant question on Peter's mind for many, many years. No matter which way he turned, life threw him curve balls. He tried to adjust. He tried to find the groove. Sometimes he thought he was right where he was supposed to be, but when he swung the bat, all he heard was "Strike three! You're out!"

Peter Tan was the product of post-World War II frugality. His parents knew what it meant to go without meals and hide from the enemy. They knew the reality of hand-me-down clothes and bomb shelters. As a result, they hammered home the lesson of choosing a profession that provided abundant income. Their children would not become dependent factory workers. They would be doctors, lawyers and dentists.

When Peter graduated from the University of Birmingham School of Dentistry, they were overjoyed. His graduation fulfilled one of their lifelong dreams. Peter was happy too, but not because he was a dentist. He was relieved to be finished with the sheer boredom and joyless applications of a school that filled cavities, examined gums and straightened teeth. By the time he graduated, he knew that this was not the life for him. The problem, however, was staring him in the face. He just spent eight years preparing for a career he didn't want. Now what should he do?

He realized that the best times he had at school were with his friends; they were the only thing that made those days bearable. Staring at teeth and grinding molars was quite different from engaging and passionately discussing all sorts of issues with compatriots. In that context, opening one's mouth was thoroughly delightful.

Following a period of serious evaluation after graduation, he acknowledged the pull of his heart strings. Dentistry was a great income but it didn't capture his passion. After considerable reflection, he decided to go where his heart was leading. He became a Pastor. How he looked forward to expounding the Bible each Sunday! But as the senior pastor of a premier Baptist church, he found he was also saddled with administrative duties.

Once again, the nightmare of activities not born from his soul became the bulk of his responsibilities. Except for those times in the pulpit, he was miserable. The daily grind of budget meetings, staff conflicts, worship organization and financial planning just left him cold. As he struggled with the expectations of his congregation, suddenly things got much worse. His wife was diagnosed with cancer.

"Lord," Peter prayed, "I have followed you and trusted you in all these decisions. I still believe that this is the way You want me to go. But now I don't understand. Why is this happening to me?"

The doctors said the prognosis was good. Angie seemed to be responding well. Six months after her surgery for breast cancer, after the horror and emotional devastation of mastectomy, Peter and Angie thought she was going to survive. Then it all came back. On one checkup, she looked great. But a month later, the cancer had spread through her lymph nodes and liver. Although she seemed to be at peace, he was devastated. His church held all night prayer vigils for his beloved wife. But in a few weeks, she was gone.

Work is part of what it means to discover ourselves. It is the junction of personal design and social connection.

The agony of her death accentuated his frustration with the operational tasks of being a pastor. He became more and more aware of the need to be engaged in what he loved – and to put aside those things that interfered and disrupted that passion. He concluded that even though he loved a part of the role as pastor, the rest was still sapping the life from him. He had to give it up.

But how could he leave behind another career?

Soon after his decision to leave the church, he was invited to speak at a conference. The experience was exhilarating. His ability to engage the audience, his desire to share with them and their enthusiastic response fed the passion of his soul. Now he knew where his talent really found expression. He gave his notice to the puzzled congregation and looked for opportunities to become a featured speaker.

Today Peter is an itinerant preacher/speaker, traveling from one hurting group to another. His message punctuated by real life experience resonated in a way he never could have duplicated. Passion returned with his usefulness to others. Over the years, he began to understand. He was born to be a channel of grace and mercy, a servant to others who had known suffering himself. He became himself and experienced the joy of being.

Work must be resurrected from the grave of expediency to the new life of an essential manifestation of who we were born to be.

"Dwell as near as possible to the channel in which your life flows" said Henry David Thoreau. That helps us see what work should be. David Whyte draws on William Blake's genius when he says, work is "to feel that what we do is right for ourselves and good for the world at exactly the same time."[1] Work is part of what it means to discover ourselves. It is the junction of personal design and social connection. To work is to be human. But most of us rarely experience the confluence of doing what is right for ourselves and good for the world at the same time. Most of us labor rather than

---

[1] David Whyte, Crossing the Unknown Sea: Work as a Pilgrimage of Identity (Riverhead Books, New York) 2001, p.4.

work. We do what we must while we hope for the opportunity to do what we love. The greatest tragedy of contemporary human resources is the mismatch of talent and career. We have been operating with the wrong normal. We have been seduced into believing that work is about the production of money, a means to an end, not the end in itself. This must change. Work must be resurrected from the grave of expediency to the new life of an essential manifestation of who we were born to be. Now we know that being human is being optimally productive in our true calling. And now, at last, we have discovered how to determine what that calling really is. The wrong normal is about to be replaced.

Chapter 2

Stumbling On The Solution

How We Uncovered The Principles Governing Human Productivity

What makes a person tick? Why do some people seem to succeed and others fail no matter how hard they try? Are we just putty shaped by training and circumstances, or are we pre-disposed to function best in a certain way?

These were the nagging questions that haunted John Samuel, sending him on a restless search for answers. His vocational career as a corporate recruiter only heightened and accentuated these issues. Every time he helped a company select a manager or an executive, he yearned for some definitive insights about how the person would perform. Will this accountant be better doing internal audit or chasing payments from retailers who need to be pressured all the time? Will this engineer be good at handling breakdown maintenance or will he be better at preventive maintenance? Will this manager lift the division to new heights or will he be someone who maintains the norm? How can we know who is best for any given job?

Man had split the atom, but human design was still a mystery.

Each candidate's résumé was always replete and bursting

with facts – where they studied, what they studied, where they worked, what they accomplished – but there was nothing that pointed to the individual's passion, what they loved to do and what they did best. In short, there were no indications where you could bet on them. Most of the time, knowing how they would perform was nothing but a guess. But predicting performance was what each supervisor or employer would ask John. "Tell us more about what he will do best", "How can we be sure she will deliver given her lack of hands-on experience in managing a team?" "Will he perform better if he's allowed to function independently or if he's harnessed to a team?"

Psychometric tools reveal slices in the life of a person, not pictures of the whole person's productive orientation.

The questions were vitally important but the candidates' résumés did not provide any definitive answers. John's training in psychology didn't help much either. He attempted to secure some conclusive answers by using the psychometric tools from his training, but most of those tools only compounded his problems. While purporting to provide answers, most of the tests were vague and suggestive at best and misleading at worst. They were slices in the life of a person, not pictures of the whole person's productive orientation.

He began to experience the same frustration that most psychologists in the marketplace know too well: The frustration of inconclusive answers – the lack of definitive inputs that will make life easier for an employer and fulfilling for the employee. Man had split the atom, but human design was still a mystery. "Don't expect to be sure about people. Trial and error may throw some light on performance, but still be wary" was his typical advice. "People are putty. You can change them into whatever you want. They only need more training". In the 80's and the 90's, training was the human resources mantra. "We can change anybody. Just give us enough time to train them." But the more the professionals pushed training, the more money spent on employee classes and seminars, the less productive increase we saw. Something was wrong with the model, but no one knew what.

John saw first-hand the limited efficacy of training. People were all hyped up after a training program but sooner than later it simply wore off. Those training dollars had no lasting value despite the best intentions of top trainers. Apparently, training had a limited impact on making people change their "default settings".

John noticed that the best predictor of human behavior was past performance. It made more sense to study how the individual performed over a period of time. His passion for clinical and observable evidence led him to design a process in which he and his team would interview at least 12 to 15 individuals who had known or worked with the candidate. These individuals included friends, colleagues, former colleagues, supervisors, family members and others. The interviews were conducted face-to-face whenever possible because John observed the candid feedback from these people was at its best when he talked directly to them.

Training had a limited impact on making people change their "default settings"

John and his team would fan out across the country, meeting a diverse range of individuals who knew the candidate, asking probing questions about their experiences with him. "Did he enjoy working by himself or was he always keen to work in tandem with others?" "What do his subordinates remember him for?" "What were some of the strengths he demonstrated that impressed you?" "What were some of his limitations or blind spots?" "What did he do with his greatest passion?"

The data began to tell a story – a compelling story of a consistency and constancy in the way the individual behaved, performed and delivered results. It was all available and out there if only someone took the time and effort to collect and collate it.

John's clients were impressed to say the least. The definitive data they had about the individual candidate helped them make some informed decisions about where they could position the employee to ensure he or she performed well. If enough effort was taken to research the past of an individual, second guessing what they would deliver best was not necessary.

But John knew that even though the data they gathered was substantive and revealed a consistent recurring pattern, it did not paint the complete picture. It provided glimpses of specific employment situations. For example, John could now tell whether the individual worked best when allowed to handle a task independently or if it was important for him to be a part of the team. The symptomatic evidence was definitive but the foundation of the proclivity for behavior and performance still remained unclear.

To John's delight the haze was beginning to lift. It became increasingly clear that if the 'symptomatic evidence' was studied more carefully and rigorously, there was a defining architecture behind it. A chance meeting with a friend in a conference introduced him to a model related to Success Factor Analysis, which was the life work of Bernard Haldane[1].

Success Factor Analysis required the individual to document life activities in which the individual experienced success and satisfaction. Typical research questions in this model were "What did you do well?" and "What did you enjoy doing?". This process helped John to see with greater clarity the truth about a recurring pattern of behaviors and productive capabilities resident in a person. The very factors he and his team in the earlier years would painstakingly collate about individuals were the ones this model set out to establish about the person. The similarities were obvious. "How does the individual prefer to be managed?" "What kind of capabilities does he consistently demonstrate at work?" "Does he like to manage people?" Success Factor Analysis seemed to complete the circle.

A human resource consulting firm using a model related to Success Factor Analysis invited John to become a Partner and a member of the Board. He continued his human resource consulting work using this framework. But after a few years of fervently championing this model, he was still left with nagging unanswered questions. The process provided a repertoire of well-defined classifications of behavior but the validity of the conclusions increasingly became an issue for John. The behaviors demonstrated by the client were not consistent with the description of the analysis. John struggled with this inconsistency and took it up with his partners but the discussion led nowhere and John chose to leave this firm. He gave up his position on the board and committed himself to his true passion – researching the design and architecture that is innate and

[1] Bernard Haldane, www.dependablestrengths.com

resident in the individual.

The years in the wilderness began. He had an established consulting practice using this model. It seemed suicidal to abandon the business. But the call to research and investigation simply did not go away. The urge and pull seemed to become stronger and stronger. Supporting himself and his wife Ruth with some savings and a meager practice, he invested countless hours searching for that elusive design that seemed to be resident in the individual. Talking to John in those days revealed a person who was passionate and determined to uncover the design he was certain existed but did not know where to find it.

> *John invested countless hours searching for that elusive design that seemed to be resident in the individual.*

His friends found him sincere and earnest but they thought he was wasting his time looking for the "hidden treasure" of human resources. The suggestions and confrontations were always the same. "It just doesn't exist. It all depends on the model you start with. There isn't any absolute here." "Why did you give up such a prestigious partnership to pursue research? You need to just get with the program and make money." His income was minimal and sporadic. His career looked like talent being wasted. John had his doubts too. Is this quest worthwhile? Is this going anywhere? Should he just give up? But silencing the urge to plod on proved to be harder than he expected. It seemed easier to relentlessly push ahead despite the financial deprivation and the attendant social stigma. His quest for the 'holy grail' seemed foolhardy, but he was soon to discover it was worthwhile after all.

Apart from his professional passion, John had other enduring interests. Visiting inmates every week in prison and doing a Bible study for them was a deeply fulfilling part of his life. His Bible studies called for an intense study of the Scripture. His passion for the Bible led him to read and study the Scriptures in Hebrew and Greek. During one preparation study, he stumbled on an otherwise obscure passage which seemed to unlock the secret he so desperately sought. It was a eureka moment. The passage made some passing reference to gifts and capabilities naturally inherent in every individual – a passage of Scripture that millions had 'spiritualized', dismissing its consequences about ordinary, innate human behavior.

John found a seminal truth in this passage which would easily have been obscured if he had not invested thousands of hours interviewing and analyzing over 8000 people the previous twenty years. All of a sudden, the pieces all began to fall into place. Finally it made sense. Just like design patterns in nature, there was a definitive design in the way people were wired and constituted. Just like design in the animal kingdom, differentiating vertebrate and invertebrates, or the plant kingdom distinguishing between shrubs, trees and herbs, there seemed to be an enduring design about how people are created. Breathless and excited, he realized he had found the grail – the architecture that governed human resource design.

> *Just like design patterns in nature, there was a definitive design in the way people were wired and constituted.*

And the rest, as they say, is history! His consulting work provided the platform to continue his research in the different continents and among various racial groups. The evidence became clearer and clearer. The reality that there is a definitive architecture to human design and that an individual's proclivities and productivity are defined more by that architecture than any other single factor, was at last empirical and demonstrable. He began to use this framework in a variety of applications including hiring for companies, helping individuals in midlife crisis make vocational transitions, team building for CEO's and their teams and helping college graduates make informed vocational choices.

A revolution in human resource management began.

To further his passion for research, John gave up running and managing a for-profit consulting outfit and started a not-for-profit research and teaching foundation. As the number of people he diagnosed increased, the clarity about this architecture strengthened. It was now possible to have predictive insights about the passion resident in an individual that could be deployed in a role that engaged their 'sweet spot' - where they were energized and were able to perform beyond expectations. The architecture was as defining as the human blood types.

This body of knowledge is called the Zone Phenomenon or technically the "Optimal Productive Function" naturally embedded in an individual. Helping individuals to secure a definitive understanding about what they are naturally endowed to do best became possible and plausible. A revolution in human resource management began.

"Living In Your Zone" was no longer a cliché. John began to see how it could become an intentional reality if and when insights about an individual's Zone became foundational and pivotal to a person's life engagements.

Chapter 3

Myth Busting: Cleaning Up the Trash

Contemporary human resource management is generally a reflection of the philosophy of the culture. Our Western culture is based in a Greek conception of the world. In Greek thought, the emphasis is on analysis – breaking larger events and processes into smaller components. Rather than looking at the whole as an organic unit, Greek thinking seeks the smallest common denominator behind the whole. This pursuit is an expression of the need to control events and things in the world. In Greek thought, if I understand how things work and what they are made of, I can influence and control the ways that the world functions.

When it comes to human resources, this fundamental way of looking at the world means that I study human behavior as component parts of a larger entity called a person. I look for common patterns among all people, hoping to find similarities that will allow me to predict and control behavior. I use investigative methods similar to the ones found in chemistry or physics – repeatable experiments or tests with control groups so that I can isolate certain behavior patterns. Understanding work becomes a function of understanding all the actions, attitudes and tools needed to produce a desired result. Since I emphasize the similar elements of any particular task, the pieces that go into accomplishing that task are interchangeable with any other similar piece. In other words, the person is just one replaceable part of the total picture. There is nothing essentially special about this human part. It is only one component among many. Sadly, the emphasis is on the productive task, not on the person doing the task.

Sadly, the emphasis is on the productive task, not on the person doing the task.

With a philosophy based in the analysis of interchangeable parts, examination of personality types or traits is viewed as another predictive challenge. Psychometric testing becomes a method for explaining what is common among individuals in relation to particular tasks. Furthermore, since human components do not differ essentially from any other component of the work process (they are still combinations of energy output, cost analysis and labor issues), human beings are seen as productivity blank slates. Effort and output is just effort and output. One person's capability is essentially the same as any other human being's capability. So, all I really need to do is provide the proper training, motivation and environmental conditions in order to reach my expected output.

The fundamental assumption of contemporary human resource management is this: People are really all the same. Training and motivation makes the difference. We can make anything happen if we just put our minds to it.

This mythology needs to be exploded. It infects the entire culture, from corporate resources strategy to weight-loss claims. It is the basis of self-help thinking, hard work commitments and motivation speakers. It is the wrong normal.

The Myth of Self-help

Why can't I just read a book in order to know more about myself? If a book can describe the necessary actions and attributes of human success, why can't I just follow the formula and achieve all that I dream of? Why can't I think my way to success? In an age where self-help books are sold by the millions, any other approach to study and appreciate the unique repertoire of the strengths embedded in an individual seems out of place. Why would we suggest that this self-help approach is bankrupt when there are so many advocates for the paradigm?

Walk into any bookstore today and you will see a plethora of books offering formula approaches to life. "Seven Steps to Perfect Health" or the "The Secrets of Financial Freedom" or "How to Lose Weight and Stay Beautiful." Each new title offers a solution that only requires you to buy the book, read it, put it into practice and, presto, you made it!

While many of these books may offer suggestions and solutions that will help the man on the street with weight control, speed reading or financial management, they provide no definitive answers about how the individual can determine or appreciate his unique range of natural strengths and gifts. By their very nature, self-help books treat the similarities among people. No self-help book was ever written for just one person – you! Self-help is really the study of common conditions and solutions for all human beings. None of these books are about the completely unique individual who picks up the book and reads it. Self-help books promote the idea of self-diagnosis. But any attempt at self-diagnosis in order to define the unique range of natural talents embedded in an individual is likely to be incomplete at best and warped at its worst. The principle "You can never see the picture when you are inside the frame!" best describes why any attempt at self-diagnosis will prove both futile and pointless. You are not the common man. You are unique and understanding your uniqueness is the key to unlocking your passion. But self-help won't help. It's about the big "us", not the particular "you."

> *The principle "You can never see the picture when you are inside the frame!" best describes why any attempt at self-diagnosis will prove both futile and pointless.*

Any successful attempt to define and determine an individual's unique talent architecture will require the support and assistance of another person to objectively provide empirical facts and data. This is why self-help will NEVER work when an individual is searching for those critical answers about what he will do best and where he will be most energized. Uncovering the natural, embedded talent architecture within the individual requires a careful examination of the person's unique history by an objective outsider. In other words, trying to examine yourself is like trying to do open-heart surgery on yourself. Theoretically, it's possible, but don't try it at home!

THE MYTH ABOUT HARD WORK

Another myth that people often cherish is that hard work and determination will make up for lack of knowledge about their natural strengths and gifts. They assume that as long as you put your heart and soul into something, you are guaranteed success. They look around and see those successful sportsmen or the millionaire business men and their intense drive and determination to succeed by hard work and they assume they just need to work harder and they will make it. When things don't turn out the way they expect, they blame their lack of intensity or their misunderstanding of the success secret. They renew their efforts, usually finding more and more frustration along the way. Laboring under the false assumption that all men are equally capable to achieving a goal, they never question the issue of innate talent. Their commitment to a false view of equality prevents them from realizing that although all men are created uniquely capable, all men are not created equal. The truth is that every person is unique, not the same as everyone else. No two people are capable of doing the same job in exactly the same way. Every one brings a unique set of talent and skills to the task and the failure to recognize this leads us to believe that performance is only a function of effort and training. As Bernard Haldane profoundly expressed it "Because you are unique, there is something you are better at than anybody else."[2]

> *"Because you are unique, there is something you are better at than anybody else."*
> *– Bernard Haldane*

Our research shows people succeed best when their hard work and determination is in the area of their natural strengths and talents. And conversely, when people are determined to be successful in an area that is out of sync with their natural strengths and gifts, they experience burn out and failure every single time.

Hard work is foundational for success. Edison toiled almost a hundred times to get the right combination of material to

---

[2] Bernard Haldane, www.dependablestrengths.com

make the filament in the light bulb work. If you had watched Michelangelo paint the ceiling of the Sistine chapel hanging almost upside down, calling that hard work is an understatement. Seeing Tiger Woods practice his golf swings over and over again proves that hard work is not just an option for success. But the key question is whether your hard work is in alignment with your natural and inherent strengths.[3]

When hard work becomes the mantra of an individual's life, driving extended periods of labor, handling roles that are not in sync with natural strengths, the individual is more susceptible to terminal illnesses like cancer and Alzheimer's disease, depression and stress-related illnesses. Most cancer researchers agree that prolonged stress in a person's life often can contribute to the onset of this mysterious disease. And hard work outside the area of your natural strengths insures that stress becomes the norm of your existence.

THE MYTH ABOUT MOTIVATION

Another favorite alternate for a strengths-based concept of work is the aphrodisiac of motivation. Enthusiastic books or talks by motivational speakers attempt to imbue individuals with a new-found sense of confidence, making them feel anything is possible.

Common in corporate sales and marketing departments, an inspirational message or a challenge by the manager is often the only way to make the sales happen. For some managers the opportunity to sit down with poor performers and instill in them a can-do spirit is almost obsessive. Their reliance on being able to influence and impact the person may at times show

---

[3] Consider the analysis of Malcolm Gladwell, *Outliers*, confirming this insight.

momentary results and improved performance – but too often it is not sustainable. Next year the process must be repeated, with new slogans, new motivations and new emphasis. The sales hype doesn't last.

Motivational talks and books that challenge an individual's sense of inertia are good and sometimes necessary, but they are a poor substitute for a life of self-sustaining energy in one's Zone. Our research shows that when individuals engage their natural strengths for the better part of a work day, they are likely to experience a rush of energy which is both refreshing and enables them to perform more effectively. We acknowledge this self-sustaining energy in musicians and athletes. We know when we are witnessing a performance that feeds on itself. But we don't ask, "Why isn't my work self-sustaining?"

Our research shows that when individuals engage their natural strengths for the better part of a work day, they are likely to experience a rush of energy which is both refreshing and enables them to perform more effectively.

THE MYTH ABOUT THE EFFICACY OF PSYCHOMETRICS

For the last hundred years, psychometrics had been the holy grail of deciphering the presence or absence of behavioral traits, intelligence and other personality factors. It is used to determine whether a student qualifies for college or an employee gets a promotion or is hired or fired. Built on the two pillars of validity and reliability, psychometrics attempts to prove or disprove the presence of behavioral characteristics specific to an individual. By correlating each response to a battery of previous responses, it purports to identify the presence or lack of abilities resident in an individual. Consequently, for the last fifty years industry has relied on psychometric testing to determine if an individual has qualities that match or suit expected work roles.

The basic design flaw in psychometrics is the inordinate reliance on an individual's response to multiple-choice questions to ascertain traits specific to an individual. When the individual is the one who is making these choices, he is exercising a self judgment that makes the conclusion subjective and therefore biased. While psychometrics attempts to over ride this subjectivity by asking several different questions which query the same idea in order to eliminate subjectivity, <u>self judgments muddy the waters as long as the individual is the principal source of providing data</u> – a flaw which often results in distorted and contorted conclusions about the individual.

The basic design flaw in psychometrics is the inordinate reliance on an individual's response to multiple-choice questions to ascertain traits specific to an individual.

As we pointed out earlier, "You cannot see the picture when you are inside the frame." This principal is fundamental to any objective assessment of the traits resident in an individual. Verifiable and empirical data about an individual can only be secured when multiple sources apart from the individual provide data that will corroborate and confirm the truth about the person.

Psychometrics fail in their attempt to study observed lived

phenomenon because the phenomena about the uniqueness of an individual cannot be validated by a process that relies on self analysis. Phenomenological research always requires multiple sources of data – and the best sources are often outside the individual.

Another major flaw in psychometrics is the substantial cultural conditioning that affects the questions. What makes sense to an American may mean something else to an Asian. Aspects about life that are the norm for the European can be unclear or confusing to an American. Once again, the process depends entirely on the self-judgment of the one taking the test.

> *Phenomenological research always requires multiple sources of data – and the best sources are often outside the individual.*

If self help, motivation, hard work and psychometric testing are not the answer, what is? Is there any definitive way we can truly determine the unique talent architecture of an individual?

CHAPTER 4

THE DNA OF OPTIMAL PRODUCTIVITY

The Question of a Lifetime

Is it possible to determine where you will be optimally productive during the entire course of your lifetime? Can we identify your sustainable competitive and economic advantage? Can we determine in advance where you will enjoy the greatest economic premium and the most personal fulfillment at the same time? In other words, can we have clarity and specificity about where you will make the most contribution and most difference to the world over a lifetime of endeavor? Is it now possible to determine the place where work is right for you and good for the world at the same time?

Our research and analysis produces an unqualified "Yes". The answers to these related questions are not found in categorized personality profiles or psychometric propensities. In spite of their continued use and popularity, psychometric tests provide only a snapshot of the real you. They are part of the wrong normal; the tendency to fit you to the job rather than fitting the job to you. But there is another way. Specific and definitive lifetime optimal productivity can be revealed, targeted

Knowing how you are wired enables you to work according to your passion code.

and understood. Furthermore, we have discovered that there are innate 'absolutes' that define where the individual will be most productive. Engaging these 'absolutes' has direct, positive benefits and consequences for career or vocational decisions. Knowing how you are wired enables you to work according to your passion code.

These innate 'absolutes' are verifiable and demonstrable. While education, training, experience, culture, family upbringing, ethnic background and opportunity can enhance these absolutes, they nevertheless remain consistent and endure during the lifetime of an individual. They are embedded in the person, not merely added to the person. They are your optimal productivity DNA.

A career or vocation built on the bedrock of these innate absolutes will enable the individual to deliver results, exceed expectations and ensure excellence because it engages what is 'hard-wired' in the individual.

The productive contribution of an individual during a lifetime is a direct function of how much he or she gives expression to these innately embedded specific absolutes. These absolutes represent the built-in capacity to function and contribute at the highest level. This capacity is the source of inexhaustible energy and effectiveness embedded in each individual. It is the single most important causal factor of optimal productivity and sustained satisfaction achieved by any individual.

This phenomenon underscores the reality of how each individual life can be optimized by uncovering, developing and deploying the specific absolutes that are innately embedded in the person. A career or vocation built on the bedrock of these innate absolutes will enable the individual to deliver results, exceed expectations and ensure excellence because it engages what is 'hard-wired' in the individual.

THE MECHANICS OF THIS PHENOMENON

The Optimal Productive Function of an individual can be uncovered and defined through a phenomenological diagnosis of the prior 'productive' engagements of the individual. This is not a multiple choice, fill in the circle method. It is an interactive and reflective process that carefully examines the uniqueness of the individual, not the common categories of multiple individuals or groups. Embedded absolutes can be uncovered by using a diagnostic methodology that interactively investigates the collected evidence where the individual has consistently demonstrated optimal effectiveness. The diagnostic procedure begins by asking the candidate to relate a few activities and engagements in the past where he or she made a significant contribution in value and benefit to others. The focus of the diagnostic method is not on the personal results to the individual but rather on the external consequences to others.

The focus of the diagnostic method is not on the personal results to the individual but rather on the external consequences to others.

The diagnosis continues by examining these data points in a one-on-one session with the person. The analysis is validated by collating data secured from several key individuals who have significantly interfaced with the participant.

The absolutes that emerge through this interactive diagnosis are compiled and presented in the form of a Productivity Compass. This compass points to the 'definitive range' where the individual is innately equipped and best fitted to contribute and function.

Our research reveals the presence of two key absolutes embedded in any individual. These two innate absolutes are foundational and pivotal to sustained productivity.

Absolute No. 1 – Natural Productive Zone:

The first innate absolute embedded in any individual is defined best as the person's Natural Productive Zone (or simply Natural Zone). This absolute defines where the individual is driven to function optimally and what kind of activities or engagements energizes him or her the most. There are seven distinct natural Zones. Our research demonstrates that each individual will function optimally in only one of these seven Zones during his or her lifetime.

Each Zone is distinct, definitive and deterministic. It provides the individual with a specific scope or range of engagements where he or she will thrive, cutting across a variety of careers, vocations and jobs. It does not define the best-suited job or profession as much as it determines the best-suited kind of engagements, outcomes and results.

Each Zone is distinct, definitive and deterministic.

For example, one of the seven Zones is the dynamic range of "teach-impart-explain". Those innately endowed with this Zone will thrive in engagements that require the final outcome or result to focus on teaching, transferring or imparting some knowledge, understanding or skill to others. Similarly, those innately endowed with the Zone "build-marshal-establish" will be energized and gravitate toward roles and responsibilities where the outcome or result requires leadership, management of personnel and the demand for decisive planning and action.

The more the individual engages in activities that are in alignment with the natural Zone, the more he or she experiences an effectiveness and proficiency that is exceptional and sustaining.

Each of these seven distinct Zones represents productive roles, engagements and activities that the individual is innately best fitted and equipped to do. The Zone defines the outcomes and results that the individual will gravitate toward in order to find purpose, fulfillment and meaning in life. The more the individual engages in activities that are in alignment with the natural Zone, the more he or she experiences an effectiveness and proficiency that is exceptional and sustaining. Attempts to function in a Zone that is not innate, native and natural to the individual will result in sub-optimal performance, stress, discouragement and burnout.

Like the definitive nature of blood types where every individual belongs to either blood group A, B, O or AB, the natural Zone of an individual is definitive and unchangeable. There is no evidence to support the suggestion that the definitive Zone can be altered or fundamentally changed as a result of education, socialization, specialized training or any other extrinsic factor. While individuals can develop and enhance capabilities within the area of the natural Zone, they cannot substitute or replace their definitive natural Zone with another Zone of their choice or preference.

Absolute No. 2 – Key Aptitude:

The second innate absolute embedded in each individual is best defined as the Key Aptitude (or Key Driver). An aptitude defines a natural capacity for effectiveness, expressed as an action, abstraction or emotion. It is what the individual is naturally 'fitted with', 'suitable for', 'constituted with', 'appropriate for' or 'created for'. It is not an acquired skill or competence. It is much more like a key driver behind the person's ability to successfully accomplish tasks. It's a person's default operating method.

The innate Key Aptitude embedded in each individual represents the most uncanny and potent 'tool' or 'weapon' of the individual. It is the 'sharpest knife' in the individual's arsenal, allowing him to optimally function in the natural Zone. The expression of this innate aptitude allows the individual to demonstrate exceptional prowess and acumen. It is the default methodology that the individual falls back on when required to handle or accomplish any critical task or mission. The more an individual gives expression to the Key Aptitude, the more the individual demonstrates a vigor and vitality that is continuously renewed and replenished. This becomes a perennial source of inexhaustible energy. Understanding and engaging the Key Aptitude with the Natural

Zone allows an individual to become a <u>self-renewing source of energy</u> and production. This is the experience often described with words like "I could do this for hours and never get tired."

Any individual may demonstrate a range of productive capabilities, but the Key Aptitude is by far the most critical and potent. Defining the individual's Key Aptitude is pivotal to determining the person's competitive edge. It helps define what he or she is exceptionally good at doing. The engagement of the individual's Key Aptitude allows the individual an edge and effectiveness that is unique, specific and exclusive to the individual. The combination of Key Aptitude and Natural Zone provide an individual with significant economic leverage and advantage. In other words, when a person understands and employs the Key Aptitude and Natural Zone combination, he or she will stand above others, exhibit significantly higher productivity and be able to command a higher premium in economic return while, at the same time, find fulfillment in being a benefit to others.

The engagement of the individual's Key Aptitude allows the individual an edge and effectiveness that is unique, specific and exclusive to the individual.

Similar to the Natural Zone, the Key Aptitude innately embedded in the individual remains fundamentally unchanged during a lifetime, essentially consistent and not altered by external factors or environmental changes.

An individual's <u>Key Aptitude falls into one of these</u> three major categories – <u>Action</u>, <u>Abstraction</u> or <u>Emotion</u>. For example, the innate aptitude to organize and classify resources or assets can be categorized as an action-aptitude. The innate aptitude to conceptualize or mentally process data or facts will fall under the abstraction-aptitude. The innate aptitude to influence, empathize or persuade is in the emotion-aptitude category.

Each aptitude has a productive and contributive value. An individual will be optimally productive in any given task or endeavor if the activity substantially engages and requires the expression of the individual's Key Aptitude as a critical need or requirement to discharge that responsibility. Therefore, deploying individuals in roles that demand optimal engagement of their Key Aptitude is an essential pre-requisite for personal optimized productivity. Furthermore, engaging people in roles aligned with their Natural Zones and Key Aptitudes provides the greatest source of productivity for the company, organization or society. If you want maximum performance, you must discover and use the Natural Zone and Key Aptitude of your people. Unless you consciously take this step, you will inhibit potential productivity, discourage maximum effort and reduce results.

Engaging people in roles aligned with their Natural Zones and Key Aptitudes provides the greatest source of productivity for the company.

Empirical Validity:

This phenomenon is consistent, demonstrable and can be validated through the productive life-engagements of any individual. Through an on-going process of 'Productivity Tracking', research shows that the optimal productive engagements of any individual will be within the scope and specifics of the Productivity Compass.

Similar to time-lapsed photography, 'Productivity Tracking' entails systematically documenting the 'peak productive performance' engagements and expressions of an individual over a finite period of time. Collating the data secured from several of these engagements provides the hard evidence revealing the consistency of the Natural Zone and the specific Key Aptitude. Research also demonstrates that the individual will be able to sustain satisfaction and fulfillment in any work, career or vocation only when he or she is engaged in roles and responsibilities that require consistent and substantive expression of his or her innate Natural Zone and Key Aptitude. Our tracking of 'peak productive performance' in the life of an individual reveals that as long as the critical deliverables in the job or role give expression and engage the innate 'absolutes' (Natural Zone and Key Aptitude), the individual is able to sustain optimal productive performance. Whenever the role is changed and the individual is required to deliver critical results that do not engage or give expression to innate absolutes, the performance of the individual becomes sub-optimal at its best - and more often than not, results in unresolved frustration, fatigue and burnout.

What Does This Discovery Mean for Human Resources?

Given the definitive and enduring nature of the data that the Natural Zone and the Key Aptitude provide, there are several benefits that an individual and the organization will be able to secure when the individual's Productivity Compass has been well defined.

Career Pathing: The Productivity Compass provides the person with inputs that are both foundational and pivotal for career pathing. It enables individuals to chart a course and navigate the twists and turns in a career with specificity and certainty by recognizing where their "edge" and effectiveness will be maximized, guaranteed and sustained during their lifetime.

The Productivity Compass provides them with both the direction and the range of engagements that will tap into their innately endowed advantages.

Optimal Productivity: Individuals equipped with a Productivity Compass analysis will be able to focus on those specific roles and engagements that provide optimal expression within their 'absolutes'. This eliminates the needless trial and error process for discovering where the individual will make the most contribution. The Productivity Compass enables them to sift through the variety of opportunities that present themselves in a job or career and be selective to be effective.

Sustaining Career Satisfaction: By using the Productivity Compass, individuals are able to define where they can attain, and sustain career satisfaction. The Productivity Compass provides them with both the direction and the range of engagements that will tap into their innately endowed advantages. Individuals who enjoy consistent career satisfaction are able to demonstrate confidence, emotional well-being and get the job done with finesse and excellence.

Energy and Effectiveness: The Productivity Compass helps individuals identify where they will be able to demonstrate an uncanny edge and effectiveness in a job, career or mission. It provides definitive data about where the person will be able to tap into an inexhaustible source of intrinsic energy that will sustain passion and guarantee satisfaction. The individual's 'center of gravity' will be maximized when the person consistently engages the Natural Zone and Key Aptitude, resulting in clarity, confidence and passion in handling and executing the given responsibility.

Educational Growth & Development: An individual's educational growth and development will yield optimal results only when the education (or any specialized training) acquired builds on the bedrock of innately endowed absolutes. The Productivity Compass will equip the individual with clarity about the kind of training and education that will enhance innate capabilities and broaden the scope and productive contribution in an organization or societal context. Training without regard to the Productivity Compass is pointless. It treats all individuals as if they were clones. Effective training is stretching the individual in the area of their innate talent and potential.

The Productivity Compass enables the individual to be clear and specific about where they can contribute best on a team.

Effective Team Work: Finding one's niche on a team is critical to the success of any collective effort. Rather than take on a role or responsibility based on a situational need, the individual who operates on the basis of the Productivity Compass will be able to be specific about where he or she can contribute optimally. This avoids any unrealistic team expectations that can lead to needless frustration and low morale. The Productivity Compass enables the individual to be clear and specific about where they can contribute best on a team.

The Productivity Compass becomes a dynamic tool in the hands of CEO's, HR managers and the supervisor.

Economic Value: An individual's optimal economic value is a direct function of how much he or she has developed and deployed the innate absolutes. An individual's economic value represents the economic premium of an individual in a market economy. It defines where the individual's contribution will be most sought after and where he or she can excel, sustain passion and be effective in delivery. Developing and deploying these innate absolutes insures their economic value is recognized and rewarded.

Return on Investment (ROI): The organization will secure its maximum ROI on human capital only when the individuals are deployed in roles that will give optimal expression to their Natural Zone and Key Aptitude. The Productivity Compass becomes a dynamic tool in the hands of CEO's, HR managers and the supervisor, allowing deployment of personnel in ways that will engage the individual and also achieve corporate objectives. It also helps the supervisor to have clarity about who will perform best in a critical task, project or mission.

Data for Deployment: Whenever the organization needs to deploy personnel to achieve a critical mission, data relating to the innate absolutes in an individual becomes critical. More than any acquired experience or training, the individual's innate absolutes play a pivotal role in determining sustained success or effectiveness. On the other hand, deploying people when their innate absolutes are not compatible to the task at hand will lead to burnout, emotional fatigue and sub-optimal results.

Data for Development: The Productivity Compass will yield valuable insights for individual development. It provides inputs and direction to HR for the kind of specialized training and skills that will enable the individual to become most proficient and capable.

A few consequences of not defining the Productivity Compass of an individual:

Sub-optimal Performance: When individuals engage in roles that do not completely give expression to their innate absolutes, our research shows that despite best effort, determination or intent, the outcomes and results will be mediocre and sub-optimal. We have observed this when individuals are moved from a job where they were peak performers to a different role outside of their Natural Zone. Despite their best efforts, they proved to be under-performers. The phenomenon of "The Peter Principle" demonstrates the efficacy of this insight. People will not perform optimally merely because they are promoted, trained or otherwise equipped for the job. The job must engage their Natural Zone and Key Aptitude to sustain passion and effectiveness.

The job must engage their Natural Zone and Key Aptitude to sustain passion and effectiveness.

Burn out: When an individual is forced to engage in roles and responsibilities that do not give expression to their Natural Zone and Key Aptitude, research shows that over a period of time the individual will experience observable symptoms of burnout. This is more marked with those in mid-career than with those who are starting out in their careers. Symptoms of burn out include physical fatigue, inertia, minimal motivation, lack of focus and little or no stamina to complete the task.

Depression: When an individual is denied expression of the innate absolutes on a long-term basis, the risk and probability in succumbing to depression is relatively high. We have observed that individuals demonstrate a lack of focus and disinterest in handling their role or executing their responsibility despite incentives and inducements to perform whenever the Natural Zone and Key Aptitude are not substantially engaged. Technically, depression is defined as a psychoneurotic or psychotic disorder marked especially by sadness, inactivity, difficulty in concentration, feelings of dejection, hopelessness and sometimes suicidal tendencies. Many of these symptoms are observable when the individual's innate absolutes are not given expression for an extended period of time.

Terminal Illness: Current research suggests that when the individual pursues a job, career or vocation that is not in line with the innate absolutes for an extended period of time, there is increased risk of cancer, Alzheimer's and other kinds of terminal illness. While the evidence is still anecdotal, it is increasingly clear that such an occurrence should not be surprising given the severity and the intensity with which individuals push themselves to be successful. When such pursuits span several years, the physical immunity of the human body is reduced. While there may be several factors including diet, environment and genetic endowment that play a role in these kinds of physical impairments, we have observed that the lack of engagement of the individual's Natural Zone and Key Aptitude for a substantial period of time during one's working life may be one of the primary causal factors for the onset of terminal illness.

CONCLUSIONS

Uncovering the innate Natural Zone and Key Aptitude (which constitutes the Productivity Compass) of an individual is the most important information to define and develop the productive and contributive value of the individual during a lifetime.

This Optimal Productive Function embedded in an individual is a naturally occurring phenomenon that can be harnessed to enable each person to make a positive, useful and economically viable contribution to a nation, society or organization. It is a win-win option that affirms the intrinsic productive capabilities of an individual and also provides the society or organization with unique contributions that will enrich and enhance its social, intellectual and economic worth and value.

If the reality of this phenomenon becomes evident to those engaged in the deployment and development of human capital, it will significantly alter the way we educate, hire, assign, promote, train, appraise, manage and invest in people.

If the reality of this phenomenon becomes evident to those engaged in the deployment and development of human capital, it will significantly alter the way we educate, hire, assign, promote, train, appraise, manage and invest in people. But if this reality is ignored, neglected or even denied, the consequential impact on any knowledge-based economy will be both damaging and far-reaching. Stifling innate talent is tantamount to emasculating the specific productive function embedded in any individual. It is damage to the heart and soul of human progress and development.

Chapter 5

The Productivity Genome – The double helix of work

Was work always toil and labor? Can work ever be food for the soul? Why has work become so much drudgery and something to be avoided if possible?

Work not birthed by passion is often nothing more than forced labor. For work to be passionate it has to be supported by the bedrock of one's natural strengths and innate productive capabilities. If work is to be fulfilling and beneficial, it has to engage the soul.

If work is to be fulfilling and beneficial, it has to engage the soul.

Consider the activities of an eagle. It loves to soar – because it was designed to fly high, seek prey and swoop with majestic grace as the master of its domain. It hardly complains about soaring every day – neither does it dread the labor of spotting its prey with precision from 3000 feet above the ground. Just so, human beings were designed to be like the eagle – to experience the delight of doing exactly what they were designed to do.

The ancients may have known the secret about soul-feeding work more than we realize. In the Hebrew world view, the expression for work and worship apparently had the same root word. Work was a form of worship, something to be reverentially delightful. The Hebrew word, avodah, means both to work and to serve, and in particular to serve the Creator. The ancients viewed work as an expression of worship because it was soul-engaging – something to look forward to each day.

Work which does not engage the soul is not merely stressful, it is dubious and questionable. It drains and depletes the individual. On the other hand, work that engages a person's natural talents and strengths energizes the individual. In many ways, work that engages one's talent design defies physics. In the physical world, expending energy depletes reserves. But when an individual engages his natural strengths, the more he expends energy the more he feels energized. There is a perennial rush of sustainable energy that is exhilarating and deeply fulfilling to the person who is "in the Zone".

"Psyche" is the Greek word for soul. Hence psychology means the study of the soul.

Consider the difference between water drawn from a water tank and water drawn from an artesian well. In a water tank, the water is depleted each time it is used. On the other hand, water drawn from a natural fountain will continue to be replenished. Similarly when an individual taps into his natural strengths, he connects with a source of inexhaustible energy that seems to be endless and deeply refreshing.

Ironically, a great deal of psychological theory seems to have missed this point all together. An examination of the etymology of the word 'psychology' reveals where it may have veered off course. Psychology is the combination of the Greek words psyche and logos. Psyche is the Greek word for soul. Hence psychology means the study of the soul but ask any psychologist to tell you about the soul and you will only hear a lot about the mind.

The industrial revolution in the last 150 years did not help either. In past centuries, products were hand-made and customized. Whether it was an exquisite carpet or crafting silverware, production required more than mere skills. It was a passionate expression of an individual whose work was soul-engaging. Nothing in history contributed to the advancement of civilization or science or commerce that did not engage the soul and passion of an individual. One can hardly imagine a Stradivarius violin made by machine or a Rembrandt painted by numbers.

A tragic outcome of pervasive industrialization was that it is now possible to mass produce goods without engaging one's soul - and become rich in the process. The motivation for work shifted from the product to the profit. People were increasingly required to work in roles that made them a cog in the machine. They were becoming more like the machines they used day after day. We now recognize that mechanization removes us from who we really are, but we don't know how to stop it.

Most of us recognize the disparity between our passion code and our ordinary labor. Many authors and researchers have attempted to steer us toward closing this gap. But our research shows that there are only two possibilities for truly resolving this problem. The first is happenstance. Some people "fall"

Nothing in history contributed to the advancement of civilization or science or commerce that did not engage the soul and passion of an individual.

into the right occupation through a serendipitous confluence of talent and circumstance. These are the lucky few. They don't seem to know exactly how it happened, and they can't quite explain it when they are asked, but they recognize (and so do we) that they are blessed to be doing what they were born to do. While they offer us hope, they provide no road map. They have what we want, but they don't know how they got it.

The other possibility is acquiring the knowledge of your Productivity Compass - through the diagnostics of Natural Zones and Key Aptitudes. When an individual intentionally seeks to engage in work that is energizing and deeply fulfilling day after day, it will require an awareness of this 'productivity genome'. Learning this secret by trial and error usually leads to trials and errors. But grasping this information about the repertoire of natural strengths and productive capabilities, unlocks the 'double helix' of one's talent design. Then choices become deliberate and success becomes a matter of execution, not elimination.

An individual's productivity genome is not a set of learned skills. It is an innate, inherent and naturally occurring phenomenon. It is a part of the individual's natural make up. The person does not have choice about his or her Natural Zone. Every person is endowed with a talent code. It is in the expression of this talent code that the person finds the secret to exceptional performance and passionate work engagement.

Our research demonstrates that each person exhibits only one of the seven Natural Productive Zone. This one specific Zone comprehensively captures the range of productive capabilities resident in the individual. The seven Zones represent the naturally occurring talent architecture distributed across all humanity. The study of an individual's Zone provides the most comprehensive understanding of the person. It provides definitive inputs about the person's proclivities and pre-disposition. Consequently the person's natural strengths and blindspots are no longer a mystery. Making sense of why an individual gravitates to some activities and recoils from others is finally explainable.

The study of an individual's Zone provides the most comprehensive understanding of the person. It provides definitive inputs about the person's proclivities and pre-disposition.

As an example, consider Sir Isaac Newton, a paradigm case of Zone 1. Growing up in the countryside in England, he was required by his family to take care of sheep. The family noticed the number of sheep returning at the end of the day was fewer than those that had set out in the morning. They soon discovered that young Newton was so engrossed in thought that he would completely forget about the sheep and often did not realize that some of them had wandered away. An appreciation of Newton's Zone would explain this convincingly. As a Zone 1, Newton's natural passion was to comprehend any naturally occurring phenomenon and to exhaustively think through ideas and concepts that would explain what otherwise was mysterious and unclear. Newton wasn't lost in thought because he was lazy or inattentive. He was lost in thought because he was at the place he was most at home.

Zone 2's on the other hand tend to be consummate executioners of a given task. One example might be a waiter who delights you with his service at a restaurant. The capacity to ensure the task gets done, to wait on another and carry out the given requirement is the passion of a Zone 2.

Have you ever wondered why some of your high school teachers made a lesson come alive and others seemed to put you to sleep? Have you experienced the mystery of sitting under a teacher who could teach you skills or knowledge in such a way that you grasped it with ease? That is what a Zone 3 is designed to do best. Zone 3 people are the natural teachers. They are clear, logical and coherent. They excel in documenting facts. They are motivated by teaching the same information to new students no matter how many times they have covered it in the past. They are enthusiastic scribes and historians.

These Zones also represent the governing principles that define human productivity.

A Zone 4 is an encourager. This is the person you go to when you are ready to give up. They are the natural enablers who delight in supporting you and making you feel stronger. You can count on them to stand by you when you want someone to hold your hand or need a shoulder to cry on.

Have you encountered someone who seems to be a natural manager of money or people. They can naturally compartmentalize, putting resources in the proper structure with ease. These are Zone 5's. They are good at growing and developing an available resource, whether it is money or people or opportunities.

Zone 6's are the quintessential entrepreneurs who love to marshal people to accomplish a mission. They are team-builders or organizational developers. They can be directive and demanding, driving people to achieve goals that they consider important.

Zone 7s are the project-driven specialists who are consummate problem-solvers. They are quick to size up the apparent need and expend themselves to meet it. They are good at bringing intensity and focus to accomplish a goal and will doggedly stick with it until they bring it to closure.

These 7 Zones represent the basic talent architecture naturally embedded in any person. These Zones also represent the governing principles that define human productivity. Our research confirms that every single individual's talent architecture is captured in one of the 7 Zones.

Navel gazing or looking into the mirror are not the best means to uncover one's Zone. It requires the objective analysis and collation of data by an individual other than the subject. No burst of sudden insight will get the person to embrace or celebrate their unique Zone. Diagnosis requires the investigative skills of a competent analyst who will piece together the different parts of the picture to help define what may otherwise never be clear to the individual.

No burst of sudden insight will get the person to embrace or celebrate their unique Zone.

Understanding your Zone is never realized in isolation nor is it a motivation of arrogance. Your Zone's usefulness is at its best when it works in tandem with other Zones to accomplish the desired results. The genius of a Zone 6 is pointless if he does not have other Zones who will contribute their strengths to accomplish a goal that requires a variety of talents. Similarly the strengths of a Zone 4 to encourage and enable another will be useless without the contribution of a Zone 7 who can bring things to closure. Each Zone has a purpose. No Zone is sufficient in itself. In the human community, all the Zones must work in tandem to bring about truly satisfying and productive results.

Each Zone is amoral. It is a capacity, not a set of behaviors. Therefore, operating with a Natural Zone can be aimed at either good or evil. We will develop this insight in the following chapters. Furthermore, no Zone is inherently better or worse than any other Zone. Blood type A is not "better" than Blood Type O. They are just the naturally occurring distinctives found in blood. So it is with Zones. The Zone defines the person's potential for passion and scope of usefulness. But it does not determine an individual's character or morality. On the other hand an individual's character and morality will determine how they choose to use their Zone functionality. For example, in our work with prison inmates, we have yet to encounter a single convict who did not use their natural strengths and gifts to commit a crime – a classic example of talents used for the wrong purpose.

The Zone defines the person's potential for passion and scope of usefulness. But it does not determine an individual's character or morality.

The person's Zone defines where they will secure the best economic advantage. This is an insight that needs to be factored into every career choice. We have observed that individuals who are in roles that engage and optimize their strengths often are the last to be laid off in any organization. There is an X-factor about them that makes their supervisors want to keep them if at all possible. On the other hand, those whose value to the organization is more tied to their skills than their passion are often dispensable. They can be easily replaced by those who are available at a lower cost.

We now realize that understanding one's Zone is not a luxury but a necessity. It is not an option but an obligation – an obligation each individual owes himself. When individuals are grounded in the definitive knowledge about who they really are – and who they are not - they are confident and comfortable with themselves. They will be more discriminating and less likely to jump at any opportunity that presents itself. They not only understand their sweet spot, they intentionally leverage it. They live to work – they don't work to live.

> *Understaning one's Zone is not a luxury but a necessity. It is not an option but an obligation – an obligation each individual owes himself.*

CHAPTER 6

THE ZONE PHENOMENON

Uncovering the Natural Laws that
Govern Human Productivity

As explained in the earlier chapters, the Zone embedded in each individual is a naturally occurring phenomenon. It is the innate, inherent and intrinsic architecture that defines where an individual will be most productive and fulfilled. It captures the essence of a person – it throws light on where the person can sustain passion, where they can give and give and give - and not run dry. Till an individual uncovers their Zone they will be a mystery to themselves – and at times a menace to others!

Till an individual uncovers their Zone they will be mystery to themselves – and at times a menace to others!

The following 7 chapters detailing each Zone are designed to initiate your understanding and appreciation of this fascinating phenomenon – the reality of a Productive Zone, naturally occurring in each person. These chapters do not exhaustively describe all the implications and applications of each Zone – they are only meant to serve as an appetizer and help you to begin to savor the awesome repertoire of innate abilities embedded in each Zone.

Each Zone constitutes a spectrum of natural productive abilities that establish the parameters of the 'sweet spot' of the individual. Technically we define these as Zone DNA's. (This incidentally is not an acronym for 'deoxyribonucliec-acid'!) Zone DNA's represent the **Definitive Natural Abilities** embedded in each Zone. These DNA's represent the 'energy buttons' resident in each respective Zone. When these energy buttons are engaged, the individual taps into an inexhaustible source of intrinsic energy which when harnessed helps the person to produce extraordinary results and demonstrate exceptional performance.

To help appreciate the DNA's resident in each Zone, we have highlighted several real-life case studies gleaned from our work with hundreds of professional clients. These case studies are in italics and will help you see how these DNA's play out in the arena of your job and career – and in also every other aspect of your life - social, marital, and family life included.

We have not only highlighted the DNA's (which represent the strengths) embedded in each Zone, we have also indicated some of the blindspots or limitations that are native to each Zone. It is critical to understand that each Zone has its strengths – and intrinsic limitations.

It is critical to understand that each Zone has its strengths – and intrinsic limitations.

This knowledge helps each person to appreciate they do not have it all! The more an individual understands their Zone the more they will be able to admit and accept their need for others to make up for their lack for what otherwise is required to get the job done well.

As highlighted earlier, you are likely to attempt to second-guess your Zone as you read through these 7 Zone chapters. Speculating what could be your Zone can be an entertaining arm-chair exercise – but a dangerous one if you want to invest your life or make career choices based on some speculative conclusions. We have in chapter 15 explained why it would be necessary for you to solicit professional help to objectively diagnose your Zone – doing your own open heart surgery is theoretically possible – but certainly not advisable! Don't try this at home!

> *Doing your own open heart surgery is theoretically possible – but certainly not advisable! Don't try this at home!*

CHAPTER 7

ZONE 1 –
DRIVEN TO UNCOVER
THE LATENT AND HIDDEN

Zone 1's see the world differently. In ancient times, these people were called prophets, but that doesn't mean they were fortunetellers or crystal-ball gazers. Prophets were those people who saw beneath the surface, beneath the ordinary activities of life. They unpacked, investigated and revealed deeper levels of reality in ways that often changed how the rest of us understood the world

You have probably never observed a sixteen-year-old chase a beam of light, but it might not surprise you to learn that Einstein imagined, at the age of 16, chasing photons. That little bit of imagination eventually led to the development of the theory of relativity – and the way we understood the construction of the universe changed forever. Our common picture of Einstein is a wild-haired man who thought about the deepest structures of the cosmos, the epitome of genius. But Einstein's behaviors are no different than many other Zone 1's. They are consumed by an incessant need to unravel and understand.

Zone 1's are consumed by an incessant need to unravel and understand.

Did you know Freud spent 4 weeks intensely searching for the testicles of eels before he moved his research into the fields of neurology and psychoanalysis? The passion to find out and make sense of what is not obvious clearly runs through Freud's life. Whether applied to the unconscious or the anatomy of an eel, the passion remained the same.

Karl Marx, born in a long line of rabbis, grew up observing the plight of the exploited workers. He researched extensively the reality of a factory worker as "an appendage" of a machine. His social-political views grew from a passion, and talent, to investigate, explain and elaborate.

These are some examples of Zone 1's who shaped and made history. Zone 1's are those for whom life is all about searching, exploring and unraveling what otherwise is hidden or latent or obscure.

Throughout the history of civilization Zone 1's have always been at the cutting edge of exploration or research. Whether it was Peter Drucker or a Shaman in an African village, they excelled in peering into the unknown. Looking for fundamental answers, they searched and studied the obvious and the obscure in order to understand something deeper.

Take the case of another famous Zone 1, David Livingstone, who sought the source of the river Nile. Doggedly he roamed the length of the continent traversing where no European had set foot before. As a young man he loved to roam the country side examining the flora and fauna that simply captivated him. The African continent opened up and the colonization of Africa began due in great part to Livingstone's explorations.

Zone 1's are designed to look at the big picture and examine the fundamentals of any phenomenon that interests them. They have an eye for the empirical and its implications for life or for society at large. They are eager to uncover the form or structure or the design of anything – and will relentlessly pursue it until they arrive at some credible answers.

Zone 1 Distinctives

Expository Excellence: William Wilberforce's compelling expositions in Parliament against slave trade was so fascinating that James Boswell described him as "a small shrimp" when he began to speak, but "the shrimp grew bigger and bigger and became like a whale" as he used his natural talent in exposition. The capacity to exposit is an innate capacity embedded in Zone One. People in this Zone have incredible gifts for oration and persuasion, gifts that often exceed professional training.

> *Zone 1's are eager to uncover the form or structure or the design of anything – and will relentlessly pursue it until they arrive at some credible answers.*

Unraveling Complexity: Before 1909, major surgical operations which required transfusions of blood were extremely risky and often - if not always - ended in disastrous complications. In 1909 Karl Landsteiner proved the existence of the four blood groups, revolutionizing surgical operations. He was able to unravel the complexity by his dogged determination to understand the phenomena.

Research Driven: Did you know Charles Darwin spent five

years aboard HMS Beagle systematically collecting species and examining their form and structure? His faculty for observation was legendary and untiring. For years he collected a variety of details relating to birds and animals, discussing his findings with other naturalists looking for commonality and design in the structure of living beings. Researching is another embedded capacity in Zone 1 people. They exhibit a natural curiosity and look for data and details that otherwise may not interest others.

Defining Phenomenon: Both Copernicus and Galileo are classic examples of the Zone 1 behavior of defining phenomenon. Copernicus' careful study of planetary motions and his heliocentric model of the universe led to succinct definitions of how the planets in the universe moved – definitions which flew in the face of wisdom cherished by the Catholic Church at that time. A few years later, Galileo, inventor of the telescope, added to this body of knowledge by defining the phenomenon of planetary motions with such exceptional clarity and precision that it shook the foundations of accepted knowledge – and resulted in his house arrest for the rest of his life. Zone 1's are adept at studying a phenomenon and formulating definitions that help capture the essence of an otherwise complex idea.

Propounding Ideology: Karl Marx's ideology was an antidote to the corrupt capitalists in Europe who reduced the worker to a mere appendage of a machine. His passion to help the worker fulfill "his potentialities as a human" was not merely fanciful. It was based on a study and comprehensive grasp of the dynamics of the exploitation that was the norm in the early

Zone 1's exhibit a natural curiosity and look for data and details that otherwise may not interest others.

years of industrialization. Marx was funded by his friend Engel who provided for him from his own income because Marx could hardly make ends meet. But his passion to "unite the workers of the world" and his dreams of a utopian society where productivity was the means for common good rather than a benefit to a select few, kept him tirelessly writing and propounding his ideology until the end of his life. Often an ideology – good or bad – sparked and stoked by the passion of a Zone 1 can blaze its way into history reducing much of whatever is in front of it into rubble.

> *Often an ideology – good or bad – sparked and stoked by the passion of a Zone 1 can blaze its way into history reducing much of whatever is in front of it into rubble.*

Exploring Frontiers: Another fascinating distinctive of a Zone 1 is the passion to explore the unknown. Did you know that Christopher Columbus believed with certitude that he had discovered the 'Indies' when in fact he had only stumbled upon the Caribbean islands? He undertook the voyages with a determination to find a sea route to India to gain access to the spice trade with Asia. His mastery of using the compass was legendary. As much as he set his sights on discovering new lands and trade routes on behalf of King Ferdinand and Queen Isabella of Spain, his drive to explore was almost compulsive. He had wrongly calculated the distance between Spain and Asia but nonetheless was determined to look for new frontiers and unknown regions. Zone One people have a natural strength to explore the unexplored.

Theorizing Ideas: A quick look at history will reveal how Zone 1s have always been at the forefront of theoretical ideas and concepts. From Aristotle to Einstein, postulating a conceptual framework of knowledge has been their domain. While a theory is only a plausible explanation for a phenomenon, Zone 1's have always excelled in making a science of what may not always be provable or demonstrable at that time. Technically, a theory is a coherent group of general propositions used as principles of explanation for a class of phenomenon. Zone 1's are the masters of theory development. Take the case of Sigmund Freud who when at the end of his life was challenged about his psychoanalytical framework, purportedly replied "It's only a theory"!

ZONE 1 LIMITATIONS

Every Zone has blindspots. These are simply areas outside of the normal purview of the person in the Zone; areas if left without correction can easily undercut the gains made by operating within the Zone.

What are some of the common blindspots of Zone 1's? Where do they often stumble and fumble? What kind of people would they need to surround themselves with to complement what they intrinsically lack?

Focused on the Big Picture: Zone 1's can be so consumed by the big picture that they can leave out details which otherwise must be factored in. There is an infamous story about Newton who is credited with designing a cat door. When his cat had a litter, he made another door for the kittens. When his friends asked him why the kittens couldn't have used the same door as the mother, he sheepishly confessed that 'it did not occur to him'. Zone 1's can be so driven by a mission that they can miss the details that will make their endeavor more efficient and effective.

Appear Dogmatic: Zone 1's are susceptible to dogmatism. Until their conclusions are subjected to the rigor of objective validation, they will need to tread slowly with their theories - and wherever necessary temper them. Given their uncanny perception of a phenomenon and their ability to see a "new" reality, they often can be ahead of their time. Since they believe that the world should conform to their vision of it, they can insist on agreement and conformity regardless of the consequences. For Zone 1's to be truly effective, they will need the wisdom and insights of other Zones to balance their perspectives.

Zone 1's can be so driven by a mission that they can miss the details that will make their endevor more efficient and effective.

Function in Isolation: Zone 1's often function independently and can be a one-man-institution. Lord Mountbatten, the viceroy of colonial India, once credited Gandhi for single handedly stopping the blood shed between Hindus and Muslims at Calcutta in the 1945, which his army of 10,000 men could not accomplish. At the same time, Gandhi's fervor and drive for an independent India pushed the political release of Britain's control far sooner than India was prepared to accept.

CHAPTER 8

ZONE 2 – MASTERFUL EXECUTOR

Do you know the best waitress in town? In all probability you were interacting with a Zone 2. Did you observe how she waited on you? Her sharp observation about what you needed and her uncanny sense of what to serve next may have delighted you. Her unobtrusive movements and attention to detail is likely to have facilitated the important conversation you had with your friend over the meal. But you will not have recognized her service at the time. The fact that she was an expert in facilitation simply faded into the background. You wouldn't realize her vital role in your life unless you reflected on what made everything "click."

Have you watched a Master of Ceremonies delight an audience with quick wit and repartee? Flawlessly engaging the audience, the time rushed by making the long evening feel like it was only a few minutes.

These two common examples show us that Zone 2's are the consummate performers who can delight an audience, execute a task and ensure they provide the personalized care an individual needs or desires.

Zone 2's are the consummate performers who can delight an audience, execute a task and ensure they provide the personalised care an individual needs or desires.

ZONE 2 DISTINCTIVES

Task Driven: *Rodney was an admiral in the Navy and enjoyed his job so much that his family often saw little of him. He was a classic example of someone who would not give up when given a task until he got it done. For example, whenever there was a Navy celebration or an event for officers and their families, Rodney was the go-to guy. He would go out of his way to organize, plan and ensure every aspect of the event was taken care of. His colleagues could never understand why as a senior officer he was taking upon himself much of what the junior officers and others could handle.*

They did not understand that Rodney as a Zone 2 was wired to be task driven – often going the extra mile to make sure the task got completed.

Operational Focus: *Jane was an intern in a bakery in Japan. She was on a student exchange program and her job was to work the ovens. Most of the earlier interns had endured this demanding role, requiring coordination of several ovens at the same time. Many of them chose to either reduce the speed of their operations or to work each of the ovens, one at a time. But to the supervisor's delight and amazement, Jane loved to work the ovens seamlessly and ensure everything was operating at maximum capacity.*

Zone 2's are wired to be task driven – often going the extra mile to make sure the task got completed.

Zone 2's are consummate in handling such an operation. When the objectives are well defined and the deliverables are clear, Zone 2's can step in and make an operation flow smoothly and efficiently.

Examples taken from real case studies are recounted in italics.

Protocol Conscious: *Jim's parents always wondered why he was so careful to make sure he was never late for school while his sister was habitually tardy and needed several reminders before she was ready to get out of the door. And Jim was so good with following procedures. He never needed to be reminded about his do's and don'ts. It seemed natural to him to identify himself with the interests of the organization. He personified protocol. He was a living demonstration of a well behaved school boy.*

Script Enabled: Amitab Bachan is an Indian movie star idolized by a billion fans. For the last 30 years, he has ruled the Indian cinema world like none other in recent history. His role in a movie almost always guaranteed it would become a box-office hit. His dexterous movements and uncanny gift to don a variety of roles, endeared him to both the rural and urban movie lover in India. When movie directors signed him to a project, they loathed to give him a script because they felt he could play his role in his own ingenious way. After all, he was the Big B (as he was affectionately known) and because he was such a consummate actor, giving him a script seemed unnecessary. In an interview with a movie magazine, Amitab reportedly mentioned that without a script, he struggled and fumbled most! Whenever the script was clear, he not only mastered his role well but also was able to creatively weave into his script whatever was necessary to make the character come alive.

Zone 2's are at their best whenever the role is well scripted and the flow is predictable. Then they become creative and productive.

Mastering Technique: Zone 2's are good at mastering the necessary techniques in order to accomplish a task or to perform a role. A technique is a skill or procedure that helps accomplish a task or endeavor effectively and efficiently. Pele, the legendary soccer player, was masterful with his footwork. His famous mantra was "practice is everything". When Pele was 15, his coach Waldemar de Brito announced to the club owners, "He will become the greatest soccer player the world has ever seen". From that moment, Pele had a penchant for formulating techniques including the famous "bicycle kick". Zone 2's are very good in perfecting techniques that appear extraordinary to others but seem effortless and natural to them.

Consummate Performers: According to his biographers, Elvis Presley was "simply at ease in front of a camera" or any audience. Zone 2's are energized in front of a crowd – and often enjoy performing to an audience. There is something almost magical about them when they are able to demonstrate or display their competence and skill. Whether it is the role of a Master of Ceremonies or a roadside singer, they come alive when they can perform or entertain a crowd. Magic Johnson earned his public nickname because his skill on the basketball floor was the epitome of "Showtime." Those who played with him often said that he played as if he had eyes in the back of his head. His one-in-a-million talent combined natural giftedness with an incredible work ethic of practice. But in front of the crowd, he was elevated above and beyond the competition.

Physical Dexterity: Michael Jackson is a classic example of a Zone 2 whose physical dexterity was legendary. His ability to perform those physically complicated dance techniques was not accidental or acquired. It was an innate ability that he mastered through countless hours of repetitive practice. Zone 2's can expend themselves in physical endeavors with almost limitless energy. As the spouse of a Zone 2 rightly explained "From the time he wakes up till he gets to bed at night, he will spin like a top – almost non-stop"! The capacity to be energized by hands-on activities is a distinctive of Zone 2's.

> *Zone 2's are good at mastering the necessary techniques in order to accomplish a task or to perform a role.*

ZONE 2 LIMITATIONS

Familiarity Breeds Comfort: Zone 2's can come across as intolerant to change. When they settle into a comfortable 'operational mode' they can resist any variation. Especially when a process works or a technique seems successful, they may avoid any change that will alter the status quo. Many times it takes a crisis or severe adverse circumstances before they are ready to embrace change.

This blind spot explains why Ann was so frustrated that her husband was 'stuck in the mud'. She was adventurous and had an appetite for new frontiers and experiences. But the more she tried to persuade her husband to make decisions that would alter their familiar routine, she simply ran into a brick wall.

Resistance to change is a blindspot that Zone 2's should recognize whenever they discover they are having difficulty pulling things together in the face of a deteriorating situation or when things are going awry.

Conflict Averse: Another interesting aspect of Zone 2's is that they can be conflict averse. Unless they are fighting for a cause or lobbying for something they consider important, Zone

2's usually avoid conflict or contexts where they have to combat or contend with another. Conversely, they tend to be compliant and are likely to take the path of least resistance.

> *For example, John was on a team where there was a lot of infighting. Every time there was a heated argument, John would quietly slip out. No one would even notice that he was not there. During a discussion concerning participation in an up-coming event, some of the team complained that the sponsor had not treated them fairly. They felt they must force the issue with the sponsor. Others felt it was not a big deal and wanted to go ahead. But John felt the angry exchanges were out of place and was extremely uncomfortable with the heated words and unkind things that were thrown at each other.*

> **Zone 2's usually avoid conflict or contexts where they have to combat or contend with another.**

Inordinate Giving: Zone 2's tend to give of themselves in a manner and measure that can drain and deplete their resources.

> *Jim was a successful sales engineer who always delighted his customers. Every time his clients wanted a demo of a new machine or whenever they needed some one to fix a break down, they could always be sure Jim would attend to them. After ten successful years in the company, his friends suggested that he cash in on the goodwill with his customers and start his own business. It seemed a good idea and Jim went for it. Everything seemed to go well until Jim discovered in spite of the good will with his customers, he was simply not able to make ends meet. On closer scrutiny he discovered that he found himself giving much more than he had contracted with them. His willingness to go the extra mile often resulted in using his*

own resources which was bad for business.

His blindspot is not uncommon to Zone 2's who tend to mismanage their resources in their zeal to delight and serve another.

CHAPTER 9

ZONE 3 – THE ARTICULATE TEACHER

Have you ever taken time to think about your best teacher at school? What made him or her special? Were you excited you finally understood that difficult subject which seemed so complex and daunting?

Can you see how a good teacher was the key factor that made a difference between knowledge and ignorance? Here's what people have had to say about their best teachers ever.

"I hated math up until then but she taught me to love it! She was so clear in her explanations and I found it easy after that to understand what she was trying to get at."

"My best ever teacher was my zoology teacher in high school. He explained everything extremely well, then checked every student's notebook to make sure the student had written down what he said and drawn the diagrams correctly."

Another said "I still remember a large part of what he taught 30 years ago – almost word for word."

"My best ever teacher was my history teacher. Yes, way back then I wrote about India and I have never forgotten what I learnt till today. She would always make the subject come alive – even the dates were easier to remember after she taught them."

Peter Drucker defined it well when he said "In teaching, we rely on the 'naturals', the ones who somehow know how to teach." Whenever you chance upon one of these natural ones to teach you something you are eager to learn, the experience is indelible and the learning is for life.

> *"In teaching, we rely on the 'naturals', the ones who somehow know how to teach." – Peter Drucker*

Zone 3's are the 'blue-blooded teachers' who demonstrate an uncanny ability to make a subject matter come alive and ignite an interest in the individual for a body of knowledge that otherwise they may not have cared for.

ZONE 3 DISTINCTIVES

Lucid Explanations: *Rachel was distraught that despite her best efforts she could not make sense of her biology classes. Her science teacher was doing his best to highlight the differences between vertebrates and invertebrates and the essential distinctives. Despite it all Rachel was struggling to grasp the details and her anxiety was building up because her exams were only 4 weeks away. She expressed her frustration to her friend Mary who lived down the street. Mary invited Rachel to come over that weekend and offered to help her. To Rachel's surprise Mary took her notes and began to teach the subject to her in a manner that made Biology seem so clear and easy. Mary graduated in Biology several years ago but once she read through Rachel's notes, she simply came alive explaining the details with such lucid clarity that in less than a few hours Rachel knew enough to be ready for the exam she had feared to take.*

---

* *Examples taken from real case studies are recounted in italics.*

When a Zone 3 highlights details pertaining to any body of knowledge, there is a lucid and succinct way in which they will express what otherwise can be confusing and unclear.

Knowledge-Driven: Zone 3's have a voracious appetite for knowledge and will seek to deepen and broaden what they know untiringly.

Mark was a HR consultant who loved to spend time in the public library. Reading and comprehending material was what made his day. He seemed to love the opportunities to train his executives about a variety of management topics. Interestingly even when hiring people for his company, every interviewee had the benefit of a half hour explanation of what he or she was good at – and also why they were not hired. Mark's capacity to quote and make reference to variety of authors and management guru's made him a 'walking encyclopedia'. And it hardly surprised anyone when Mark left HR after a few years to become a lecturer in the local community college.

Logical Expressions: *Sandra was a feisty woman who often got into trouble. The main reason she often found herself at logger heads with both family and friends was not because she was physically violent but because she would argue with such forceful logic that people found her difficult to relate to. On the other hand whenever she was teaching at the local college, her students sat spellbound at her deductive reasoning. The dry subject came to life whenever Sandra was there showing her students how to deduce the idea or see the beauty in the logic of a concept.*

> **Zone 3's have a voracious appetite for knowledge and will seek to deepen and broaden what they know untiringly.**

Because of this reason Zone 3's are also good in debate and in highlighting the pros and cons of a policy or a concept with compelling clarity.

Cogent Clarity: Zone 3's are good story tellers. They can weave a plot into a cogent story that will be both entertaining and educating. The theme of their story will flow seamlessly from one scene to another giving a comprehensive picture of the event or experience. Take the case of Jane Austin, the author of 'Pride and Prejudice'. Her stories provide a riveting plot about common place social activities and bring out life's lessons in a manner that connects and enlightens the reader. Walter Scott described Jane Austen as one with 'a talent for describing the involvements and feelings and characters of ordinary life' – the stuff story tellers and novelists are made of.

Energized Discussion: Zone 3's love debate and dialogue. The free flow and exchange of ideas stimulates them and expands the limits of their knowledge. Whether they are hosting a talk show or participating in a scholarly discussion, they excel in being able to synthesize their thoughts into forceful arguments and clear statements. The Jewish notion of Midrash captures the essence of this idea. A Midrash provides the context for dialogue, discussion, debate – which when recorded becomes an educative commentary, for future reference and review.

Keen Intellect: Zone 3's thrive in contexts which require an appetite for knowledge and intellectual comprehension. Academia is an arena where they excel because it provides ample opportunities to engage one's mental faculties.

Helen had a doctorate in mining technology from Germany and was heading a major mining project in her native country. Her colleagues found her very engaged in strategic planning, handling training for her team, writing concept papers, etc but very disconnected with hands on management of the project she was tasked to handle. Her inability to manage her team soon required her being moved to a position where she became the advisor to the Chairman of her organization – a role in which she functioned effectively because it required her to process complex data relating to the operations of the company and advise her Boss with the necessary inputs to help him make an informed decision.

Audience Sensitive: Zone 3's have an uncanny ability to understand the dynamics of how to present an idea to an audience in a manner that will be educating and stimulating. Some of the best screenplay script writers and dance choreographers are Zone 3's who can orchestrate and choreograph a scene to be visually impactful and also convey a key message or idea. Take the case of the famous dancer and choreographer Agnes de Mille whose genius was to reflect "the angst and turmoil of the characters, instead of simply focusing on the dancer's physical technique". Zone 3's have a natural ability to make the subject matter come alive and present it to an audience in a manner that will make the message indelible and fixed in one's conscious memory.

Zone 3 Limitations

Poor Managers: Zone 3's are poor managers much because they would rather facilitate the growth and understanding of a person rather than supervise or monitor their performance.

Adrian was an excellent teacher in a well known Graduate school and his classes were the most sought after by students in his College. When the Principal of the college retired after nearly two decades of being at the helm, the Board decided to offer the position to Adrian who was one of the senior most academics on staff. Adrian recalls he was hesitant and uncertain because the classroom was his first love – not the staff room where he had to roster the faculty and supervise some of the junior staff. But his wife nudged him on because becoming the Principal in an Asian college meant moving into the Principal's bungalow and with more servants and added privileges. Everything went smoothly for the first couple of years before the cracks started to show. Staff morale started to go downhill, the students became restless and within a few months the college was shut down for an interim period – only because Adrian cloistered himself in the office, refused to deal with controversial administrative issues, buried himself more and more in his books, almost like the proverbial ostrich which pretended the problem didn't exist. Predictably he had to go and later found his niche in becoming a Curriculum specialist in another Graduate school.

Structure Dependent: Zone 3's are structure dependent and function optimally when there is a structured routine.

Rose was an effective teacher as long as she was a part of a structured academic school. On account of her family circumstances which did not permit her to do full time teaching, she chose to take special tuition classes for those who needed help in learning English. Many of her friends spread the word around that she would be available to take these part time classes and soon she had a good number of students who would visit her every evening for her special tuition classes. But soon Rose found herself experiencing burn out – it was not the teaching that was wearing her down but the administrative details of scheduling the classes, collecting the fees and managing a structure that would ensure the classes followed through seamlessly. On the other hand, when the structure is available and well defined, Zone 3's function optimally and effectively.

Zone 3's are structure dependent and function optimally when there is a structured routine.

Not Hands-on: Have you ever wondered at the statement "Those who can do, those who can't teach"? There may be more truth to it than what most people would imagine. Zone 3's personify this statement in its totality.

Take the case of Raj who was one of the best choreographers in the Indian movie industry. While he failed miserably in his attempts to be a movie star, he turned out famously as a choreographer. He could teach and coach those dance moves and choreograph scenes in a way that would delight his movie Directors. But he was simply no good when he had to perform the role himself. Anita was another example of how in her role as a Production Manager, she was very good at coaching those who were reporting to her, but was very disappointing when she had to get her hands dirty to get the job done.

Zones 3's are very effective in roles which require teaching another, but can struggle and fumble when they have to be hands-on and execute the task themselves.

CHAPTER 10

ZONE 4 – THE QUINTESSENTIAL ENCOURAGER

Have you ever wondered why when you spoke to certain people, they seemed to instill in you a can-do spirit when you felt down and out? And why when the whole world seems to give up on you, they are willing to stick by you? When every one tells you that you are a failure, these people have that undying hope about your recovery?

It is most likely that you were interacting with a Zone 4. People endowed with this Zone are those you can bank on to encourage and enable you. They will ensure being there as much as you need them. You can trust them to provide for you from their resources, as long as they see there is some hope for you to bounce back.

Zone 4's are those you can bank on to encourage and enable you.

Michael had a rough childhood. His parents had divorced when he was three and much of his life was in foster homes or with relatives. His social skills at school were so bad no one wanted to be around him – at least till Roger became his classmate. Roger made sure he took intentional effort to be around Michael, stand by him when he got bullied and slowly started to help him come out of his shell. *After high school Michael went on to become a top notch software programmer for a Silicon start up firm. A few years later Roger received a beautiful card from Michael telling him that he was the only one who truly cared for him and how much he owes his success to him.*

Zone 4's are the natural encouragers and without them many a success story may never have happened.

Some Zone Distinctives

Equipping Enabler: *Sarah was a business owner who provided the community she lived in with vegetables and fresh fruit. While she made a decent profit, her real passion was helping the girls in the local high school to run a tailoring class. She loved designing and sewing and was eager to see these girls have some kind of livelihood when they graduated. So Sarah spoke to the Principal and asked if she could donate a few sewing machines and teach a weekly class for the girls as a volunteer teacher. The school agreed and Sarah was there every week teaching and coaching these girls who could not believe someone would take the time and effort to so passionately give of themselves to train and equip them. Many of the girls after they graduated went on to run their own small tailoring shops and the others became mothers who ensured they stitched clothes for all their family needs.*

Zone 4's like Sarah have a natural capacity to enable another and equip them with all that will help them to grow wings

---

*Examples taken from real case studies are recounted in italics.

and fly. They love to come alongside those who are looking for someone who will support and enable them to grow stronger.

Inspiring Coach: *Derek was a Math teacher in the local community college who took it on himself to coach the basketball team. Rain or shine he always showed up on the court to cheer and coach from the sidelines. To ensure the college physical ed teacher did not misunderstand his initiatives he made sure that he worked in tandem with the physical ed teacher and offered to share his load whenever his time permitted it. What amazed all the kids who were on the basketball team was that Derek's passion was consistent and authentic – whereas the college PE teacher was simply carrying out his duties - and doing it often perfunctorily.*

Zone 4's are natural coaches – they can push, challenge and motivate others.

Zone 4's are natural coaches – they can push, challenge and motivate others to stretch and become proficient and more competent in their area of work or expertise. You can trust them to effortlessly coach and mentor another as long as the person is eager and interested.

Motivational Support: Norman Vincent Peale created a near-cult following in the 70's and 80's teaching and challenging people to believe in themselves and in the power of positive thinking. Apart from his role as a pastor at the Marble Collegiate church in New York, he set up a practice in which he claimed much success motivating people to believe in themselves and in applying the principles of the Bible. Listening to Norman Vincent Peale preaching was an experience in itself – it was high energy, challenging and instilling a can-do spirit in the hearers. People would leave his church service fully persuaded that they can handle adversity, change their attitudes and face life better because of what they heard him say.

Even if all Zone 4's are not so high powered motivators like Norman Vincent Peale, they are very good at motivating and instilling confidence in another. They have a natural capacity to make you feel better and more confident, after you have talked or interacted with them.

Effortlessly Adaptive: One of the fascinating DNA's embedded in a Zone 4 is this natural capacity to be flexible and adaptable. This DNA represents the capacity to change, alter and ensure becoming relevant and suitable to the context. A Zone 4 is adept at adapting to a context and blending into the milieu. They can quickly adapt to a new scenario and relate to others in a manner that will be inclusive and non-threatening. This DNA helps them embrace diversity and relate effectively with people with whom they may be either non-familiar or dissimilar.

A Zone 4 is adept at adapting to a context and blending into the milieu.

Mary was good at Chinese and picked it up with an effortless ease. She was born German and even at school her facility to pick up French and Spanish was remarkable. But ever since she met a few Chinese girls at her church she was fascinated by the culture and began to take Chinese classes at the local University in Germany. Soon she not only mastered the language but her intonation and expression was like a native Chinese. People who heard her on the phone would have sworn she was a native Chinese and not a European.

What was even more interesting was when she was visiting China many were amazed at how well she adapted to the local customs and just blended in. Excepting for her blonde hair and other physical features, her manners and deportment would have passed for any Chinese woman who grew up in China.

Zone 4's can easily adapt and relate to different contexts – however diverse they may be. This natural capacity helps them to be very good in cross cultural contexts and in embracing diversity very effectively.

Collaborative Peer: Zone 4's thrive in flat non-hierarchical structures. They do well in contexts where they can relate to others as peers and equals.

Don was the head of a department in a manufacturing firm. He had a knack of connecting with the Janitor and relating to him as if he were his equal. Ever since Don became the boss, every one noticed the Janitor felt motivated and eager to do his best. Don was good at making even the most subordinate staff feel on par and a part of the team.

Don's management style always involved the entire team in decision making. He loved to have every one contribute their ideas and looked for consensus. While this at times infuriated some of the senior members of the team who wanted Don to be decisive and unilateral, Don's patience in bringing every one on board, helped the team morale to flourish and be up-beat.

> *Zone 4's love to relate to people as peers and will invest much in mutual trust.*

Zone 4's love to relate to people as peers and will invest much in mutual trust. They are very engaged when they are able to work through collaborative relationships in which each party does their part with a sense of responsibility and commitment to the common good.

Facilitative Catalyst: *Gwen was a school counselor who enjoyed mediating and bringing warring students together. She loved to sit down with both parties and give them a patient hearing. Prior to her appointment this inner city school was famous for its gang fights and racial conflicts. After she came on the scene the word spread that going to Gwen would help both parties to win – neither would lose face. The Principal was so impressed with Gwen's contribution; she recommended that Gwen help a few of the neighboring schools, which were rife with gang fights and other student problems.*

Gwen's DNA to mediate and be a facilitative catalyst found good expression in this role. She loved to see the warring and angry young people reconciled and helped them to appreciate each other better.

Zone 4's love to facilitate harmonious relationships – and in their role as catalysts they are effective in bringing about a change in an individual, team or any ineffectual context.

Adept at Handling Multi-engagements: Zone 4's are adept at handling diverse engagements concurrently or one after another. They enjoy the variety and range of activities that may be dissimilar or unrelated. They thrive in roles where they are expected to handle the multi-faceted aspects of a task or endeavor. Routine, repetitive and predictable work often bores and depletes them.

Tom was masterful in handling the role of a college Dean which required his daily interface with faculty, students, parents, civic authorities, the administrative team – and even the workers union. His predecessor who was a brilliant teacher had left this position after being hospitalized twice because the range of things to handle was simply too overwhelming. But for Tom it was energizing and brought the best out of him.

This DNA to handle multi-engagements is a natural capacity embedded in Zone 4's. They thrive when the variety is a given and they are required to handle a diverse range of activities or engagements.

Zone 4 Limitations

Unrealistic Optimism: Zone 4's tend to be perpetually optimistic. They love giving people second chances – they find it hard to give up on people they care for and want to help. Consequently they may make poor managers whenever the role requires being directive and there is need to monitor the performance of another. Zone 4's tend to give individuals on their team a long rope before they take any decisive action to let them go.

Zone 4's tend to avoid confronting another as far as possible.

Zone 4's need to temper their optimism with some factual, and realistic inputs from those who have a shared interest in the team. Otherwise they can find their decisions about poor performers to be belated, obsolete and damaging to the team.

Conflict Avoidance: Zone 4's tend to avoid confronting another as far as possible. They are conflict averse and tend to look for a harmonious and amicable way to resolve issues whenever they arise. Consequently they will tend to procrastinate difficult people decisions as far as possible. If they are in a managerial role, they will struggle with decisions related to firing employees or in confronting those who are poor performers.

Chang was an excellent HR manager in Singapore. His boss began to trust him more and more in making strategic people management decisions. When the economic downturn hit Singapore and there was a compelling need to downsize and let people go, Chang was caught in a bind. Cognitively he was clear that he had to retrench at least 1 in ten employees – but he loathed and resented his having to do it. It was gut-wrenching.

And as he struggled with it, he realized that even the employees who were asked to leave the company were not taking it so personally but he was having sleepless nights – every night - just having to handle this exercise!

Free Spirited: Another interesting downside to Zone 4's is their proclivity whenever possible, to want to be free to do what they want, when they want. They love the freedom to flit from one activity to another, even if it is not completely finished.

Jane was talented in producing jingles for radio stations. She had worked several years for a few large radio stations and her friends encouraged her to become a free-lancer to beat the long hours and the occasional weekend work at the radio station. She felt tempted and reasoned if only she could work just as hard for herself she could make a lot of money – and importantly enjoy her freedom.

Six months after she set up shop in her home office, she realized she was not able to meet deadlines and deliver what she had promised her clients. The challenge of doing household chores and sitting at her work desk seemed more difficult than she had imagined. And before a year was over, her productivity had dropped and her clients were not giving her repeat orders.

Zone 4's function best when they are a part of a structure and when they have flexibility within the structure.

Zone 4's function best when they are a part of a structure and when they have flexibility within the structure. Whenever they tend to be solo players or start an enterprise independently, they seem to flounder and struggle – sooner than later!

Chapter 11

Zone 5 –
The Resource Manager

Have you wondered why some people are good at managing and growing money or developing a natural resource? They seem to have a knack for being able to spot a good investment and know how to make their money or a resource grow. Also have you wondered why some people enjoy managing and coordinating people? They seem to be able to delegate and distribute work in a way that the job gets efficiently done.

These are the Zone 5's – the quintessential resource managers. They have an amazing capacity to be able to manage, apportion and develop resources. They love to contribute and provide a resource when it will help develop and nurture another. For a Zone 5, *giving* is a fulfilling endeavor – in fact the elixir of life!

For a Zone 5, giving is a fulfilling endeavor – in fact the elixir of life!

Zone 5's demonstrate several embedded capabilities ranging from managing resources, to being efficient organizers, to those having a drive to nurture and develop people or other resources. They view time, talent and treasures as resources that can be optimized and developed to accomplish something significant and to impact the world around them. Let's look at some of them in greater detail.

Zone 5 Distinctives

1. Give and Make Available: Zone 5's love to give and make available a resource that will benefit another. Whether it is making a philanthropic donation or providing someone with an opportunity which otherwise they would not have, Zone 5's are eager to provide and contribute for the benefit of another.

Jim, a young lad from Seattle, loved to collect butterflies. He had an exquisite range of those butterflies neatly arranged and preserved in several beautiful glass cases at his school biology lab. His teacher who was always impressed with Jim's ability to hunt and catch butterflies, applied on his behalf for a grant to a university. This allowed Jim to travel to South America and along with a team spend time in the Amazon collecting a range of butterflies native to that region. Jim was one of the best on the team and the butterflies he was able to collect were of such exquisite nature that the State museum requested for them. Jim was delighted to donate them after he organized them in beautiful glass cubicles in which they were exquisitely mounted for display.

Zone 5's desire to give and make available things that will benefit and be useful to another. Like in the case of Jim, donating to the museum was what made his joy complete. Collecting and acquiring those butterflies was fun, but making them available for the benefit of others was deeply fulfilling.

2. Control and Coordinate: Zone 5's have a natural flair to control and coordinate an event, a project or a team. They love to assign roles and ensure the harmonious functioning of a team. Collectively and systematically they will seek to accomplish

---

*Examples taken from real case studies are recounted in italics.

what needs to get done by harnessing all those who are willing and available for the task at hand.

Noor was a Production Supervisor in a Japanese TV manufacturing company in Singapore in the 70's. Her Boss was so impressed with her capacity to manage her team that soon she became the "Lead Girl" which meant she would have to assign duties to her team mates and also ensure the production targets were achieved both in terms of quantity and quality. Noor took to this role like a fish takes to water. It was a matter of time before she found herself being able to achieve production targets that were the envy of others on the production floor.

Zone 5's demonstrate a capability to control and coordinate an operation with proven ease and efficiency. Like Noor, they thrive in situations that require systematic control and supervision of an operation.

3. Organize and Structure: Zone 5's are also good at being able to organize and structure a program, event or any given resource. They are good at putting things into a flow and structure to ensure the objectives are accomplished efficiently and systematically.

Zone 5's are good at putting things into a flow and structure to ensure the objectives are accomplished efficiently and systematically.

Stan was a Swiss agronomist that enjoyed pasture management. So when his University requested him to undertake a study that would ensure the pastures in the Alps were maintained in a way that was sustainable, Stan worked on the project with gusto and much enthusiasm. What impressed his supervisor was the meticulous way in which he tabulated the findings, compared and contrasted the data with the yield from different years and made sure he was able to submit a report that was both factual and easy to comprehend. What stood out about Stan was that many of the templates he created were original and organized in way that was detailed and appropriate. Even though it was difficult to put together he enjoyed painstakingly creating a well structured and detailed report.

Zone 5's are good at bringing structure and organization to any endeavor – be it a team or an event or simply putting data together like Stan. They are naturals at taking what is undefined, disorganized or not in order and seeking to bring structure, systems and definition to make it more efficient or effective.

4. Manage to Maximize: Making sure that the resource is maximized is natural to a Zone 5. They love to manage a team or an event where the resources are optimized, fully utilized and to ensure there is no wastage.

Peter was appointed the Prefect in the Boys Dining Hall and was responsible to ensure all meals were served on time and that the food was good. Prior to Peter, the management of the Dining Hall was disappointing and the students always complained about the kitchen staff and were perpetually unhappy about the quality of the food served. The school Principal had observed how well Peter had managed the class finances and so decided to put him in charge to see if he could turn things around. In less than three months the reports he began to receive were overwhelmingly positive and encouraging.

Peter was there much of the time ensuring the cooks had the adequate supply from the kitchen store and ensured discipline during meal times. He made sure the best vendors were selected and had them submit quotes that were then displayed on the dining hall notice board. Importantly he made sure that kitchen staff functioned in a way that every item was accounted for and whenever there was excess food, Peter personally arranged for the food to be given to the School watchman and the gardener who lived on the campus leaving nothing to be wasted.

Like Peter, Zone 5's have a flair for ensuring optimal management of resources, and will either save that particular resource to be used in the future or will leave nothing to be wasted away. They will seek to extract the most from what is available and will manage to deliver the most productive returns.

Making sure that the resource is maximized is natural to a Zone 5.

5. Develop and Grow a Resource: Zone 5's will seek to develop and nurture resources in a way that will yield the best returns. Whether the resource is money or an individual, they will seek to systematically grow and develop it optimally.

Karen took on a team that was dysfunctional and performing poorly. For several reasons the earlier manager had messed up operations and the situation seemed pretty hopeless when Karen came on board. Karen began to work with the team in earnest by taking time with each of them and understanding their frustrations and disappointments regarding all that had happened. After she figured out what were the key issues, she began to initiate team building activities, which often took them outdoors and kept them engaged with her and with each other. Very soon the results began to show. The team began to become vibrant and competitive. What impressed Karen's manager was the way in which each team member was helped to grow and develop. *This was done in a way where even if they left the firm, each team member had grown to become confident and capable of delivering results that were unthinkable before Karen's intervention in their lives.*

Like Karen, Zone 5's are good at nurturing a resource and developing what otherwise may remain sub par. They will focus on growing the person or a talent or finances in way that there is systematic development over a period of time. Like the farmer who plants a seed and nurtures it till the crop comes to harvest, Zone 5's will take the necessary effort to bring to fruition what they are developing over a period of time.

6. Sort and Segregate: Another interesting natural capacity embedded in Zone 5's is their ability to sort and segregate what otherwise is a confusing mess of issues, data or physical items. They have an innate capacity to compartmentalize items and assign them to the relevant 'boxes' or compartments.

Judy was an accountant who was a natural in taking her clients' confusing and unorganized account entries and would passionately sift through the data to move those entries into the relevant expense or income columns. She seemed to be energized doing what was otherwise simply frustrating for many of her colleagues and managers. They would marvel at her passion and patience to sift through the hundreds of entries and make the postings, which tallied the balance sheet, into an impressive format of numerical elegance.

Zone 5's can sort and sift through a confusing maze of data, issues or ideas and compartmentalize them in a way where it is homogenously classified or collated. Whenever Zone 5's attempt to make a significant contribution to a person or an endeavor, they will tend to sift and segregate the details to help with both clarity and objectivity.

7. Custodian and Steward: Zone 5's are good custodians of whatever is entrusted to their charge and keeping. They will meticulously take care of what is under their charge and steward it to ensure that it is well maintained and in good order.

Take the case of Ron, who loved to buy foreclosed homes and refurbish them. He would spend hours working on those homes to make sure that when they were rented they would be cozy and comfortable for the tenants. His tenants loved him, because every time something needed to be fixed or changed, Ron was more than happy to attend to it. This puzzled his friends, who wondered why Ron was keen on doing what other home owners did not think was so critical. In fact he was once pointedly asked why it was so important for him to take such good and meticulous care of his property. Ron then explained he was simply being selfish – because he wanted his houses to be maintained in the best possible condition, so that whenever he chose to sell them, they will need no further work.

Zone 5's are good at nurturing a resource and developing what otherwise may remain sub par.

Zone 5's like Ron often will go to great lengths to maintain, preserve and ensure the care of whatever is under their charge or custody. They will continuously ensure the best possible maintenance to steward the resources they are entrusted with.

8. Spot and Invest: Zone 5's have a knack for spotting a good investment or potential for investment in an opportunity, context or person. They are naturals at identifying long term value and can quickly estimate what will give a good return in the long run.

Zone 5's tend to gravitate towards activities and engagements in which there is some return or futuristic value.

Myra was an investment manager in a finance company. Unlike her colleagues, she always seemed to know where the longer term returns would be. Her quarterly reviews did not top the list of best returns, but she was always a sure winner. Apart from doing her homework well, Myra was focused on the fundamentals more than what seemed to interest and impress her colleagues. She was willing to forego immediate gains for solid long term profit. Often her investment decisions would baffle her supervisors, because she was willing to put money on obscure funds that never seemed to be attractive to others. But her doggedness to understand the fundamentals and then go for it, even if others were not inclined, paid off almost every single time.

Zone 5's have an intrinsic capability to look for what will be a good investment in the long run – be it people, opportunities or money. Like Myra they live for the fruit of what may happen after a long time and will invest their resources in endeavors that will provide the best returns, time and time again.

Zone 5 Limitations

As with other Zones, Zone 5's have their share of blindspots, which finds expression in compulsive behaviors. Let's examine some of them.

Resource Fixation: Zone 5's tend to gravitate towards activities and engagements in which there is some return or futuristic value. They can become very resource conscious and avoid contexts or relationships in which they do not see long term growth or value.

James abruptly dropped out of the community college he was attending. His parents and friends were rather surprised – and also disappointed. James had shown much promise and potential, and they expected him to graduate with flying colors. When James' trusted friend sat him down and asked him why he abruptly stopped his course of study, James explained he found no point or purpose in investing his time in arts and philosophy classes when he was able to accomplish much by helping his dad in his training business.

It was a couple of years before James realized that if he had plodded on and completed his course of study, he could have applied for an attractive scholarship program in the local university, which required him to have graduated successfully.

Management by Control: Zone 5's tend to control the dynamics of a team or a context in a manner that can at times be stifling and restrictive to those they are supervising. They may attempt to demand details and assign roles unilaterally, which may be offensive to those they are managing.

Rose was a high school Principal who had helped turn around a dysfunctional school. While her school Board was extremely happy with her performance she had a high rate of attrition on her staff. When

Zone 5's can at times become fastidious and excessively focused on details that may not be necessary or even relevant.

one of the Board members sat with a few staff, who had expressed their desire to leave, she learned more. Rose who was otherwise a good manager, would spend hours asking the staff for details about their course work and curriculum in a way that undermined their sense of autonomy and competence.

Zone 5's, like Rose can at times become fastidious and excessively focused on details that may not be necessary or even relevant.

Acquisitive Streak: Another interesting blindspot relating to Zone 5's is their acquisitive streak. Zone 5's at times may come across as being compulsive and obsessive in the way they collect and acquire resources. Whether the resources are intangibles like skills, knowledge, education or even university credentials, or tangibles like property, money or assets, Zone 5's tend to accumulate them, often when they may not be immediately necessary.

Simon's house at times appeared like a junk yard. He loved to collect what was discarded, but had value, hoping to recycle them and sell them sometime. As an antique collector, people in his city admired his range of antique tractors and cars that he displayed every year in the County Fair. But his spouse complained she was always put off when he would bring things home without consideration for the limited space in their garage. Whenever Simon's friends gathered at his home for any social event they were always amazed at his collectibles, but also noticed how they had such limited space navigating their way to his living room.

Zone 5's tend to collect and acquire anything that they perceive to have value and long term worth.

Chapter 12

Zone 6 – Builders & Entrepreneurs

Have you noticed that some people seem passionate about making things happen and love to build from scratch? They savor the challenge of a conquest and look forward to taking the "bull by the horns". They don't shy away from confrontation if they feel someone is getting in the way of their goal. They enjoy pushing ahead.

These are Zone 6 people, the ones who are natural entrepreneurs and builders of organizations, teams or any dream they think is worth achieving. They tend to be hard drivers and can at times doggedly push (and shove!) people and resources to make things happen.

Zone 6's enjoy rallying and marshalling people to build an enterprise or wage a crusade. They are natural expansionists and love to enlarge boundaries as much as they can. Their embedded capabilities range from building an enterprise to taking a team to new heights. They can be both combative and competitive.

Some Zone Distinctives

Let's examine some of those defining strengths embedded in Zone 6's.

1. Direct and Marshal People or Teams: Zone 6's are adept at marshaling and rallying people to accomplish a mission. They tend to inspire loyalty and will attempt to fulfill a vision or project by getting people to fall in line and make sure things get done.

Zone 6's enjoy rallying and marshalling people to build an enterprise or wage a crusade.

Jeb was an Indonesian college student who enjoyed getting his classmates involved in community work. On his own, he set up a social service club that reached the rural areas surrounding his town. On a weekly basis, he took a group of students to build houses, clear the drainage or teach the village kids math or science. What impressed Jeb's professors was the ease with which he could gather a group of students around him to visit these villages once a week. They never complained about the manual labor (including those who came from rich families) and they shared a sense of purpose and mission in getting things done. In the field, Jeb acted like a circus ring master, assigning roles, moving from group to group and overseeing the whole effort. His team enjoyed pulling together and Jeb made sure everybody had a piece of the action.

Zone 6's like Jeb have a natural capacity to make things happen by directing and marshalling people to accomplish a mission or endeavor.

2. Expand and Enlarge Boundaries: *Gupta trained as a veterinarian and had a successful practice in western India. His practice provided him with adequate income, but he was very dissatisfied. His yearning to set up an entrepreneurial enterprise gnawed at him. He soon took up a pharmaceutical dealership and for the fun of it hired a couple of salespeople to visit the different*

---

*Examples taken from real case studies are recounted in italics.

veterinarians in his region introducing the range of drugs he was representing. In a few months, Gupta found himself increasingly giving more of his time to the pharmaceutical business than to his veterinary practice. He enjoyed training his salespeople, visiting difficult customers and exceeding the targets the pharmaceutical company had assigned him. The business grew so fast he started to manage dozens of salespeople who were formed into several teams. Today his dealership covers most of Western India.

Zone 6's, like Gupta, savor the challenge of extending their boundaries and expanding their domain wherever and whenever possible.

3. Command and Take Charge of a Mission: Zone 6's are good at taking charge of a context and orchestrating people to move in the desired direction. They love to step up to the plate and make sure things get done, however challenging it may be.

Cruz was a French economist who was hired by a major car manufacturer to study the risks involved in the several business interests the company had in Europe. On one occasion, one of their companies inadvertently agreed to supply a production process system without factoring in the cost for the software development. Since the deal was already made, the company stood to lose several thousand Euros. When Cruz learned about this, he offered to step in and lead a team of volunteer software programmers who would create the necessary package after office hours to avoid any financial loss to the company.

Zone 6's savor the challenge of extending their boundaries and expanding their domain wherever and whenever possible.

Though it was not in direct line with Cruz's job responsibilities, he took charge of the group who rallied behind him to create the software program within four weeks. Cruz savored every moment of his new role. He would stay up late, brainstorm with the team, ask for daily progress reports, demand answers for delays and was completely immersed in the nuts and bolts of what needed to get done. Since all the participants were volunteers, he walked the thin line between egging them on and pushing for results. His team saw his passion and enjoyed the sense of mission that he brought with him. Though they found him frequently snappy and demanding, his capacity to push them all to reach the goal brought pride to all.

Zone 6's are those who savor and relish the opportunity to command, control and drive a project or enterprise.

Zone 6's are those who savor and relish the opportunity to command, control and drive a project or enterprise. While they may invite the inputs and perspectives of those on the team, they will tend to unilaterally decide the course that needs to be taken.

4. Determined and Decisive: Zone 6's tend to move forward on a mission or project in a determined way. Their decisions also tend to be quick, pointed and decisive – and at times impulsive. Once they have their mission in the cross hairs, they will use every available resource or advantage to accomplish the goal.

Thomas was an American social worker in Haiti for several years. He gave up a comfortable career in the construction industry and, with his family, moved to Haiti to help serve the poor and needy of this impoverished nation. Thomas had a passion to build homes. Whenever funds were available, he would marshal the local villagers and build a dozen houses or more at any given time.

Thomas' wife observed that whenever he had the funds and the necessary construction material, he would launch into the effort with the gusto of Napoleon going into battle. He would be up early in the morning recruiting the local villagers, deciding who should do what and by when. Once the plans for the day were drawn up, those whom he hired soon learned about his no-nonsense approach. Like a general on the battle field, he would firmly and pointedly tell his workers to do exactly what he wanted them to do. He could change his plans during the day and would expect his workers to quickly adapt to the changes and maintain the momentum.

Like Thomas, Zone 6's will drive a team to get things done with a determination that can be infectious and dogged. Their pace and quick decision making will be charged with a sense of mission and they will expect the team to implicitly fall in line to get things done.

5. Combat and Conquer: Another interesting feature of Zone 6's is their proclivity to be hard driving and competitive. Once they are determined to win, they can display a combative side which may not always be obvious or perceived. They love to confront those who are a perceived threat to their vision or those who are poor performers and will not hesitate to engage in an adversarial relationship if it will help accomplish their vision or mission.

> *Another interesting feature of Zone 6's is their proclivity to be hard driving and competitive.*

Mary was on the school debating team and, sooner than expected, was appointed the team leader. Her school had never won *a debating competition and she was determined to make it happen. Every day after school she would meet with the group for four hours practicing and rehearsing their arguments. She would brainstorm with them about every possible counter-point the opposing teams might express. Her team members were surprised there was so much ground to cover to become robust and ready for the event.*

They soon got the idea – winning was the only option, it was expected that everything that had to be done to get there, needed to be done. Mary even dismissed a team member who was playing truant and was lukewarm in her efforts to win. When the person who was removed confronted Mary and asked to be taken back, Mary gave no ground for her decision. She was going to win no matter what.

Zone 6's are given to conquests. They are good in overpowering and prevailing over their opponents. While this helps them to achieve greatness, they can be insensitive and dispassionate in their dealings with those they work with.

6. Daring Risk-takers: Zone 6's love to push the envelope and live on the edge. Charging forward in the face of danger makes them come alive. They relish taking new ground by daring to go where most people would be afraid to tread. The routine and the predictable make them restless and they yearn for challenges that are daunting, and at times foolhardy.

Martin finished his business studies and started a food business that grew. He soon had a chain of restaurants with a swelling customer base. He was the envy of his classmates, because he made it without many losses and his capital had multiplied several times over. Every one expected Martin to continue adding restaurants to his chain and retire from the food business. To everyone's shock, Martin decided to sellout and go into the electronics business. At first his new venture progressed well but within two years Martin lost all his fortune distributing a range of products that were technically obsolete.

The banks were not confident in lending to him and many of his friends thought they had seen the last of him. They expected he would quietly retire and call it a day. But Martin persisted and after five years profitability returned. He eventually grew the business which became more successful than his earlier one. However, it was Martin's spouse who took the brunt of his perseverance. She explained that during the course of their marriage she felt as if she was on a roller coaster, never being sure what decisions he would make next. When asked to take the road that was familiar, Martin would gravitate toward other riskier options. Previous to understanding her husband's makeup as a Zone 6, she found him difficult to understand or explain.

7. **Vision-casting and Making it Happen:** Zone 6's love to envision a grand plan and will drive to make it happen. Their dreams become their passion and until they realize them, they will not rest or let go. With the tenacity of a pit-bull, Zone 6's demonstrate a drive to make their dreams become reality.

Zone 6's love to envision a grand plan and will drive to make it happen.

Romeo finished his doctorate in social studies in Denmark and wanted to work with the poor in Bombay. His friends recall how he would spend hours describing what needed to be done to alleviate poverty in the slums of Bombay and provide people with *sustainable employment. Romeo would often be found on the streets of the slums talking with the residents. He started clubs, which were managed by local people. Many before him had tried to work in these slums, but left frustrated when any significant transformation failed to occur. But Romeo changed the rules.*

Rather than initiating more welfare programs, he began influencing those at the grass root level, who in turn took ownership for development where it was most needed. Romeo's wife and friends described how he would meticulously specify what needed to be done, craft compelling proposals and persuasively recruit foundations to support his work. Romeo's friends were amazed at his ability to conceptualize a detailed plan and match it with an equivalent amount of verve and determination to accomplish his vision through others.

8. Assign and Authorize: Zone 6's love to assign and direct people to play a role or handle a function. They are also keen to empower and authorize those reporting to them. They enjoy providing those on their team with the needed resources and backing their need to accomplish the mission.

Luke was a notorious gang leader. He helped build a loyal following of inner city youth who were always available to do his bidding. The young men and women bonded to Luke because of his faithfulness to them whenever they had to pull off a crime. They were completely assured that if anything would happen to them, Luke would take care of their families. He was always watchful to give his gang members money and the authorization to call up other team members should they get into trouble. His comrades felt special and empowered when they carried out Luke's orders. Once a part of Luke's gang, a sense of identity and an esprit de corps developed that made all feel stronger and more confident.

Zone 6's like Luke have a knack for taking ordinary people and empowering them with a sense of mission, while marshaling them to accomplish something extraordinary.

> *Zone 6's have a knack for taking ordinary people and empowering them with a sense of mission, while marshaling them to accomplish something extraordinary.*

ZONE 6 LIMITATIONS

Zone 6's can also become victims of blind spots. Some of these include:

My Way Or The Highway: Zone 6's must be careful not to be unilateral and directive in the way they lead a team or manage a group. At times they can be demanding and may not accommodate dissent or divergent views. This is especially the case when they are fixated on a vision or mission they think is critical or grandiose.

Marcus was an up-and-coming entrepreneur in Beijing. His ability to spawn several successful businesses made him the envy of his friends. He would envision a business idea and then relentlessly pursue it until it became self-sustaining. His competitors feared him and his customers loved him – but his staff walked on egg shells when dealing with him. Marcus was not given to small talk and when he wanted something done, there was minimal room for discussion or debate. Mike joined the team as the Finance Director and was used to a collaborative style of relating with his boss. He loved to brainstorm ideas and then go with the best option available. To his amazement, Mike found that Marcus was willing to discuss ideas, as long as it was Marcus' decision that was finally implemented. Eventually, Mike left the team. He was simply unable to work with a boss who solicited inputs but usually ignored them when they did not support his perspective.

Foster Sycophants and Cronyism: Zone 6's can inadvertently foster cronyism in their zeal to build a cohesive team. In their drive to get things done, they may not have time or patience to seek or appreciate consensus. This facet explains Margaret Thatcher's famous comment: "My style will be conviction politics, not consensus politics!"

Mark was the head of a religious order spanning several countries that had grown in membership. While Mark was charismatic and effective in getting people to improve and transform their spiritual lives, he surrounded himself with people who never dissented or asked the hard questions. When some of the educated and mature members in the order challenged Mark to take into consideration the views and inputs of those who didn't always agree with him, he explained that he could not work with those who were not "of one spirit" with him, and he continued to avoid them. Before long, a serious leadership crisis emerged which lead to Mark's removal from his leadership role.

Rash and Impulsive: Zone 6's tend to make decisions on the spur of the moment and these decisions may not always be wholesome or balanced. They like to intuitively figure out the best available options and move forward, even if it seems foolhardy to those around them.

Napoleon's decision to abruptly attack Russia when winter

was approaching continues to perplex strategists. If he had not taken such a rash and impulsive decision, the history of Europe would have been very different.

Many Zone 6's enjoy confronting others, and often when the prospect of winning seems imminent, tend to make moves that can confound and surprise observers.

CHAPTER 13

ZONE 7 – A PASSION TO MEET NEEDS

Do you recall those friends who went out their way to care for you when you had a pressing need? Those who savored the opportunity to stand by you when you were in a crisis? Those folks who just could not sleep well until they got to the goal or brought things to closure?

These are the Zone 7's who delight in meeting immediate felt needs, who love to satisfy requirements and are driven to achieve a goal. They are the natural relief workers, those who come alive when there is a disaster or an emergency. Be it a firefighter rushing into a burning building or a surgeon doing an emergency operation - Zone 7's will step in and take charge until the problem is solved or the need has been met.

Take Mother Theresa for example. The Sisters on her team would tell you how she came alive when she was tending to the poor and dying in Calcutta. When most people looked away, overwhelmed by the hopelessness of the situation, Mother Theresa would dive in and provide solutions, even if it was just helping a sick man on the streets of Calcutta to die with dignity.

ZONE 7 DISTINCTIVES

1. Provide Aid and Meet Immediate Needs: Zone 7's look

Zone 7's thrive and are deeply fulfilled when opportunities to meet the needs of another come their way.

for opportunities through which they could meet the immediate pressing needs of another. It is hard for them to walk away from a context in which they are able to provide relief to those who sorely need it. Whether the context is a colleague who is unable to perform because of a personal crisis or a homeless family because hurricane Katrina devastated their city, Zone 7's love to provide aid and meet the needs of another.

John's family and friends always wondered why he kept adding to the number of adopted children he and his wife Mary had. John and Mary had two biological children of their own and 3 years later felt strongly about adopting Kim, who was born to an unwed mother and was up for adoption. Then year after year for 6 years in a row, John and Mary would receive calls from the state adoption agency asking if they were interested in adopting a child who desperately needed a home. Every time they would balk at the idea, but would eventually say yes, because in John's words "I could not withhold that care for a child when I was able to provide it".

Zone 7's thrive and are deeply fulfilled when opportunities to meet the needs of another come their way.

2. Trouble-shooting and Problem-solving: Zone 7's tend to be the proverbial Mr. Fixer. They come alive when there is a breakdown or if something is not working. Their interest may wane after they have fixed what is broken, but during the time they are attending to what needs to be fixed they can be extremely focused and absorbed.

---

* *Examples taken from real case studies are recounted in italics.*

Mike was a college student in Eastern Washington and was the go-to guy on Campus to fix computers. While the college had a Computer Department with paid staff, Mike served as a volunteer and did more than 80% of repairing, fixing and managing the computers of the faculty and students. What amazed people was the way Mike would 'be transformed' when fixing computers or trouble shooting a badly installed program. Whenever he had some free time, Mike was a sure fixture in the Computer Department. While the staff in the Computer Department would often drag out a seemingly unsolvable problem, Mike would step in and his colleagues would say 'we can go to sleep now'! Mike would be at it till it got done and could go for 5 hours without a break. Trouble shooting and fixing computers was not a chore for Mike – it was pure passion.

> *Zone 7's come alive when they can fix problems – be it machines, relational issues or any kind of breakdown.*

Zone 7's come alive when they can fix problems – be it machines, relational issues or any kind of breakdown.

3. Pointed and Directive: Zone 7's can be very directive and pointed when they attempt to accomplish a task or endeavor. They may unilaterally make decisions that will help with a resolution and may not always see the need to collaboratively function with another. But in their determination to get at the goal, they will be very clear and unambiguous about what must get done. Consequently, they make excellent movie directors, project leaders and event managers.

Tim was a book store manager in a small town in Ohio. Every week they would host a book club who would meet in the store foyer.

Tim would arrange the event with the zeal of a party enthusiast. While he was not very excited about the day to day operations of the store, organizing the book club meeting was the highlight of his week. He would arrange the details and have his staff execute it with the flourish of a movie director. He was obsessed with the details to flawlessly flow in sequence, like he had planned. Anything less was unacceptable. While the staff appreciated the success of the book club event week after week, they were not able to understand the fastidious emphasis Tim would place on the event, or his driving it without much collaborative inputs from those on his team.

Zone 7's delight in meeting immediate felt needs, love to satisfy requirements and are driven to achieve a goal.

4. Advise and Orient: Zone 7's are good in advising and orienting those who need to learn the ropes and are unfamiliar with the process or a system. They often will expend much effort and care to teach, demonstrate or provide the necessary inputs to help those who need 'to come up to speed'.

Matthew was the training manager for new hires in a large company in India. Whenever a new person was hired, Matthew made sure he invested substantial time to orient and provide the person will all that they needed to settle in well. While his predecessors had done the same job without much enthusiasm, Matthew took extraordinary efforts to be there for the new hires. What surprised the HR Director was the singular focus with which Matthew would contact and connect with the new hires and ensure they were ready and adjusted well in their new role. Often the new hires assumed that all the managers were as helpful as Matthew, only to discover it was not always so.

Matthew, as a Zone 7, was a natural in orienting and advising those who needed to learn whatever was required and necessary for their new context.

5. Linear and Focused: Zone 7's have an extraordinary capacity to progress towards a goal or a target in a linear and focused way. They thrive when the requirements are clearly spelled out, the objective is well defined, and they are given the freedom to get it done without too many distractions or interruptions.

Mark as a boy loved playing with Lego blocks and could spend countless hours building complex models that were the envy of his friends. When he grew up he trained as an architect and with singular focus enjoyed creating artist's models of a development or a new building. What amazed Mark's colleagues was his intensity to progressively work towards completing his artistic creation, yet his utter frustration when he was required by his boss to change tasks when in the midst of his work. While Mark's friends did not mind the interruptions (in fact some loved it!) it turned Mark's world upside down. Several unpleasant incidents later, Mark's supervisor began to appreciate the fact that it was better to let him finish what he set out to complete before he was asked to handle another task.

> *Zone 7's thrive in contexts in which they have the freedom to work at something and bring it to closure before they go on to the next task or project.*

Zone 7's thrive in contexts in which they have the freedom to work at something and bring it to closure before they go on to the next task or project.

6. Bench-mark and Standards-driven: Zone 7's have the embedded capacity to bench mark and measure up to standards. They are good at gauging the stated requirements and will strive to ensure they satisfy those requirements.

Take the case of Margaret, who worked for a global consulting company. She seemed to understand her client's issues within hours of meeting with them. As a Six Sigma consultant, she was trained to look for those aspects where processes and systems had broken down or were designed poorly. While all of Margaret's colleagues had the same kind of training that Margaret had, and some of them were longer in the business, they all agreed Margaret had an uncanny edge over others in being able to pick up what was missing or where the standards were being compromised.

Zone 7's are good at understanding the requirements or standards and will apply themselves to ensure they match those bench marks or stated grades.

7. Anticipate Scenarios: Zone 7's are good at being able to look at the different scenarios and ask the 'what if' questions. They tend to deliberate and weigh the different options and outcomes before they take the plunge. They are a natural in charting a plan of action and will prepare for any exigency that may occur, if and when things go awry.

James was a doctor in Singapore who had a reputation as the best family physician in his part of the city. Many of his patients always remembered him for being exceptionally careful in prescribing antibiotics and his high success rate of prescribing proper doses. Whenever a patient asked him to prescribe antibiotics, James would take several minutes to explain the possible scenarios that could happen when antibiotics were consumed and highlight their role in the healing process. This was a markedly different approach from other GP's in the area and James' patients always remembered him for vividly painting a picture of each possible scenario, when explaining the side effects of antibiotics and their potential impact on long term health.

8. Goal-driven and Seek Closure: Zone 7's tend to be goal driven and feel very restless until they have brought things to a closure. They loathe leaving tasks or issues unresolved.

Jean was an interior designer who loved her work and enjoyed creating beauty and value for her clients. But her husband could never understand why, after landing a project that she would stay awake endless hours designing blueprints and working on the details. Even though the project had a 6 month run time, Jean was restless until she tied all the loose ends. It was near misery for her spouse - who urged Jean to relax and take it easy. However Jean was fixated and restless until everything was completed, signed off and handed over to the real estate developer.

Zone 7's tend to struggle with projects or task which have no clear ending or time lines. They are keen to bring things to a closure and until then, can be restless and unsettled.

ZONE 7 LIMITATIONS

Tunnel Vision: While Zone 7's tend to be fixated on driving to the completion of a task or a goal. They often need time to switch or swap tasks. It usually takes a while before they can 'zoom out and zoom in'.

Florence was an efficient Secretary and someone her boss could completely rely on to get things done. Her only problem was when visitors would come for their scheduled appointment with her boss, she would often be oblivious to them, even if they were near her desk waiting to get her attention. Though she would profusely apologize, every one could see her intense focus and complete immersion in her task, made her oblivious to the world around her.

Zone 7's tend to experience this tunnel vision syndrome every time they are in flow and completely absorbed in bringing issues or tasks to closure.

Unilateral and Directive: Zone 7's can be unilateral and directive when they know what needs to get done. They may not be very collaborative or seek the inputs of those on the team.

Ram's friends could never explain why Ram behaved differently when he became the Project Lead in a software company in India. They noticed that Ram, who had been a friendly person and who enjoyed having fun with his team mates of 5 years, suddenly took on a directive and unilateral approach after being promoted project leader. When Ram's supervisor sat him down and asked why he was not collaborating with his team, Ram explained that he had worked out how the project should run and felt no need to solicit the inputs of those on his team.

Zone 7's can be unilateral and directive when they know what needs to get done.

Zone 7's need to be cognizant of this blindspot to ensure they work cohesively with those on the team.

Individualist and Specialist Role: Zone 7's are at their best when they are not required to manage many people and have the freedom to function in a specialist role.

Ted was an upcoming pharmaceutical chemist who was hired by a company before he finished his final exams. He was like all new hires taken through the usual "management trainee" route and for a few years enjoyed being a specialist researcher. After 5 years of enjoying his work, he was promoted to Departmental Manager and had the responsibility of supervising 15 executives and support staff.

Zone 7's tend to function optimally when they can play the role of a specialist, without having too much administrative responsibility.

To the surprise and anxiety of his supervisor, Ted began to slip with his deadlines, the quality of his research work dropped markedly, and his department was dissatisfied with the way he was managing. Ted's boss quickly saw what was happening and moved Ted back to a specialist role, with all the attendant benefits and compensation of a departmental manager. In less than six months Ted was back in his full form and started to deliver results that made his boss both excited and relieved.

Zone 7's tend to function optimally when they can play the role of a specialist, without having too much administrative responsibility.

CHAPTER 14

UNCOVERING YOUR PRODUCTIVITY COMPASS – ITS IMPLICATIONS & APPLICATIONS

The preceding seven chapters provide an overview of the seven naturally occurring Zones. Zones are universally present in every race, people group or culture. Because everyone operates from a particular Zone, we often miss seeing the impact of Zone behavior in every day life. We are blind to the reality of Zones because they are woven into the very fabric of living. But the truth is this: Zone behavior drives economy. People who operate from their passion are the ones who optimally build, create, perform, organize, analyze, investigate, influence, conceptualize, monitor, strategize, empathize, customize, etc. They are the ones who produce exceptional products or provide a service of excellent quality that is maximally useful to others and have a premium economic value.

But the truth is this: Zone behaviour drives economy.

The 7 Zones are a naturally occurring phenomenon that governs human productivity. Whenever individuals operate in roles that are aligned with their Zone, they demonstrate a passion and productivity that is unparalleled and seems effortless. Work energizes them – and at times becomes addictive. They bring an effectiveness and gusto to their work, which makes them adept at accomplishing what they set out to do.

As explained in Chapter 4, while Zones define an individual's overarching purpose in life, there are two other absolutes that constitute the individual's Productivity Compass. They are the Key Aptitude and the Productivity Orientation specific to an individual.

The Key Aptitude is the one pronounced and potent natural aptitude that triggers and enables the individual to function optimally in the Zone.

For example, Isaac Newton's Key Aptitude - to conceptualize and formulate a theory - helped him to accomplish his Zone 1 purpose of being a consummate researcher and expositor of a latent phenomenon. This resulted in the formulation of natural laws that governed our perception of the world for centuries.

The other absolute is the individual's Productivity Orientation. This represents the approach that an individual uses when confronting day-to-day activities. There are three distinct kinds of Productivity Orientation:

- *Action Orientation*
- *Abstraction Orientation*
- *Emotion Orientation*

In simple layman's language they can be defined as 'doing', 'thinking', and 'feeling' orientations, respectively. But they are significant in their implications, because unless the individual is engaged in tasks that are in alignment with their orientation, they will tend to be enervated and depleted rather quickly.

As highlighted earlier through an example in Newton's life, as a shepherd he would be lost in thought (abstraction orientation) in spite of the fact that he was required to manage sheep (action orientation). This incompatibility often happens in the work place where those with an action orientation end up handling abstraction roles. The result is sub-optimal performance or burn out.

Don't navigate life's journey without your Productivity Compass!

It is not in the scope of this book to define and detail how to deduce and decipher the Key Aptitude and Productivity Orientation of an individual - both of which are also naturally occurring phenomenon in the individual. As with the Zones, these absolutes are uncovered by examining where the individual made a difference to another, and was engaged and absorbed when doing so.

Uncovering the three absolutes that constitute the Productivity Compass of the individual is the single most important investment of a person's life. It will liberate and unleash the potential resident in an individual – and harness and enhance the productive contribution in every sphere of life, work, family or social involvement.

Uncovering your Productivity Compass is critical and foundational. Don't navigate life's journey without it!

Chapter 15

Rainbow Hunting

Living Happily Ever After

Michael Jordan was one of the world's greatest basketball players, but he couldn't hit the inside curve in a baseball game and he isn't one of the world's best golfers.

Bill Gates is a genius in computers, but he wouldn't do very well in a biogenetics lab.

Yo Yo Mah plays the cello better than anyone in the world, but he can't handle the blues guitar.

Warren Buffet understands stock values and financial markets, but he doesn't develop fiscal policy for the White House (although any President would benefit from his expertise and advice).

What made each of these men economic powerhouses was focus. Knowing how to leverage what you were born to do is the secret to personal economic premium. Operating within your Zone is the best way to maximize your own economic value. The key to economic success in the world is building an economy around who you are, not fitting into the work that the world prescribes for you. More

Operating within your Zone is the best way to maximize your own economic value.

often than not, those who become real economic powers are the ones who saw the world differently and decided to change the world to fit their vision, instead of conforming to the world's view of reality.

We all know how to recognize talent when it comes to athletics or music, acting or the arts, but we don't apply the same intuitive awareness when it comes to our own work. Far too often we allow circumstances and need to direct our choices in the marketplace, rather than living according to our talent code. But the intuitive understanding of talent is built into us. We can tell right away if a ball player really demonstrates excellence or a guitar player has "the touch." We know when we are in the presence of genius whether it's viewing a painting or watching a tennis match. But we rarely use that intuitive awareness on ourselves. When it comes to work, we just choose the jobs that offer the most money and the best benefits. We rarely choose our careers on the basis of our innate talent design. When it comes to work, we seem to forget who we are.

Imagine how boring the world would be if those shining examples of creative genius took the same route. Imagine what it would be like to watch a ballgame where all the players were there just for the money, or go to a concert where the musicians performed on the basis of how much they got paid. It would be depressing, discouraging and diminishing. The real "spirit" of the event would disappear. This is one of the reasons why we often find more excitement in amateur sporting events where the players perform simply because of their love for the game. Too often money begins to reshape the performance of sports professionals. We know the difference when it comes to others, so why don't we see ourselves with the same perspective?

The real reason we overlook our own talent in most job choices has more to do with the culture than Zone myopia. Our culture teaches us the wrong normal when it comes to work. Our educational system has become a product of the production-consumer culture. It is a graduated pathway toward making us fit the productive needs of the society. Far too often educational choices are determined on the basis of the hoped-for career or the suitability for the potential job market, rather than on the basis of who we are. In fact, there is no college course designed to answer the most fundamental question in life: "Who am I?" "Where will my talent be most useful?" But unless we know who we are and where my talent will be most useful to the others, any adventure into the working world will be a journey into the heart of darkness.

There is no college course designed to answer the most fundamental question in life: "Who am I" "Where will my talent be most useful?"

Malcolm Gladwell, author of The Tipping Point and Outliers: The Story of Success, summarizes our dilemma when he says:

"... autonomy, complexity, and a connection between effort and reward – are, most people agree, the three qualities that work has to have if it is to be satisfying. It is not how much money we make that ultimately makes us happy between nine and five. It's whether or not work fulfills us."[5]

"Hard work is a prison sentence only if it does not have meaning. Once it does, it becomes the kind of thing that makes you grab your wife around the waist and dance a jig."[6]

So, how do we change all this? How do we stand up against

---

[5] Malcolm Gladwell, Outliers: The Story of Success, p. 174
[6] Ibid., p.175

the enormous pressure to choose a path on the basis of monetary reward and corporate productive fit? How do we make choices based on who we are rather than who society tells us to be? How do we build an economic engine around ourselves?

Perhaps we need some insight from Gladwell's research. In his view, economic premium combines nine crucial factors:

1. Demographic timing
2. The 10,000 hour rule
3. Parental expectations and inputs
4. An economy desperate for your skills - economic timing
5. Opportunity recognized and seized
6. Cultural legacy
7. Work that is meaningful and autonomous
8. Persistence
9. Daily learning – life long pursuit

Let's look at these nine factors.

Demographic timing is beyond your control. It is a matter of how many people enter the market at a given point in history. War, recession, plagues, culturally changing technology and other macro-sociological factors play an enormous role in this. But you didn't choose when to be born. The world you entered is the "given" of your environment. In fact, it is the "given" of everyone alive during your years of work. So, recognizing its influence is important. It isn't something you can change, but it is something you can use.

However, the 10,000 Hour Rule is entirely up to you. Practice, practice, practice! Gladwell discovered that those who really shine in their chosen fields have spent at least 10,000 hours developing skills that are based on their natural talent. That's about 40 hours a week, 52 weeks a year for 5 years. In other words, what Gladwell discovered is that the people who are outstanding in their fields – and who command the highest economic premium – are those who practiced a very long time. They worked at it. They developed their natural talent with dedicated perseverance. They took what they were gifted to do and did it for a long time. No wonder they rose above the competition. Unless you love what you're doing, you won't stick with it for 10,000 hours. But if you love what you are doing, each hour brings joy – and improvement. You keep going because it is just who you are.

Knowing how to leverage what you were born to do is the secret to personal economic premium.

You're probably asking yourself the practical application question: "What steps do I take in order to gain economic advantage from this Zone paradigm?" You say to yourself, "OK, I see the truth in Gladwell's nine steps. I've done the diagnostic Compass. Now I can just apply the nine factors to my Productivity Compass and I will be successful, right?"

Unfortunately, no! Discovering your Zone and making the 10,000 hour commitment doesn't mean you're going to be a millionaire. Your Zone plus commitment might mean that you will find deep satisfaction in work, but it doesn't guarantee you will reap financial reward. Actually, building an economy around your Zone is a by-product of the alignment, not the goal.

What the Productivity Compass does is set you off in a direction. It doesn't give you an immediate financial result. It's about movement toward something, not the destination itself. But the "something" is not a carefully defined object. It isn't $1 million in the bank or a new car or a dream vacation, like those things we are often told to imagine as rewards in goal-setting exercises. Here the Compass gives direction, not destination.

Building an economy around your Zone is a by-product of the alignment, not the goal.

Once you begin to move in the direction of your maximum productivity – your Zone – then you are more likely to encounter those events and circumstances we would call serendipity, those elements of "luck" that bolster your efforts.

With a Zone framework, you will be able to recognize the opportunities that fit the Zone and you are willing to take a shot at what is offered, knowing that it brings you more in line with your direction. It's getting in sync with yourself. Each action refines who you are and intensifies the focus. You can push the envelope and take risks and be almost sure of success.

There is a difference between a "risk-minimizing" strategy and a "success" methodology. The Zone paradigm does not provide an individual with a step-by-step guarantee of success. It provides a foundation of assured serendipity.

Zone participants are not risk averse, because they realize that directed risk-taking moves them further along the direction of their ideal. They are not afraid to take a chance, because they realize that pushing forward in alignment with their Zone provides opportunity for development and refinement in what they are naturally gifted to do – and doing that will greatly increase their chances of success.

In other words, they believe that the universe is conspiring with them, because they are following the direction of what they were born to be and what they were best designed for. They push themselves in the direction of their Zone, because they know that this direction fits them into their correct place in the world.

The Zone paradigm does not provide an individual with a step-by-step guarantee of success. It provides a foundation of assured serendipity.

Chapter 16

Chasing Rainbows

How do I build an economic engine around my Zone? The best answer to this question is to consider the difference between finding the pot of gold and chasing the rainbow. Zones are about direction and purpose in life. They are not about achieving specific measures of success. While most success manuals provide you with steps for finding the pot of gold, Zones points you in the direction of life-long satisfaction. Once you know your Zone, you will know which rainbow to chase. It's the rainbow that is in the direction of your Zone.

Zones point you in the direction where your design is most in alignment with the universe and where serendipity is most likely to occur.

Chasing the rainbow means following a path that leads to greater and greater expression of your Zone – and that itself is the reward. Rather than concentrating on some set of behavioral steps to help you reach a specific goal, Zones point you in the direction where your design is most in alignment with the universe, where serendipity is most likely to occur and, consequently, where you are most likely to discover economic premium. You still have to invest the 10,000 hours. You still have to seize the opportunities. But each hour you invest and each risk you take is leading you deeper and deeper into Zone fulfillment. The very fact that you are moving in that direction provides you with an economic advantage.

Rainbows have a particularly interesting feature unlike most other observable phenomena. Every rainbow is unique to the observer. In other words, the rainbow that you see is not exactly the same one that I see, even if we are looking at the "same" rainbow. This is a result of the fact that a rainbow is a refraction of light from a particular angle. The angle of your eyes to the water vapor in the air is not precisely the same as the angle of my eyes to the same water vapor. Therefore, you do not see the "same" rainbow even though the physical phenomena we observe is the same. This is a bit like your Zone.

A Zone is the general description of the maximized design inherent in you, but the application of the Zone to your life is uniquely your own.

A Zone is the general description of the maximized design inherent in you, but the application of the Zone to your life is uniquely your own. Your rainbow isn't my rainbow, even if we both look in the same direction. So, Zone Phenomenon points you toward the observable phenomena of maximized potential, but exactly how you get there and what it will mean for you along the way is different for every human being. This is why cookie-cutter approaches to self-help and success will ultimately fail. You are not the same as everyone else. Any program or psychological diagnostic that is based on the common categories of human beings is an insufficient guide. Rainbow chasing is unique to you.

So, where do I invest my 10,000 hours? The answer is: in the direction that most embraces and enhances your talent code, your personal application of the Zone. We really can't tell you specifically what that looks like. If we could, there would be no reason for you to read any further. If all you had to do to be a millionaire was follow the steps outlined by countless books about wealth, then everyone would be a millionaire. The fact is that only the author of the book will get rich because you will buy the book. No book, no program, no step-by-step guide is the magic pill of success, because every life is unique.

Zone development does not come from thinking about your Zone. It comes from pursuing your Zone with actions.

Rainbows have another feature that is analogous to Zones. No matter how hard I attempt to chase the rainbow, it recedes from me at the same pace. That's the journey of Zone involvement. It doesn't end. The more I pursue the refinement of my Zone, the more I discover there is yet another horizon, another development of becoming me. The goal of maximized productivity recedes from me in spite of the fact that I get better and better at what I do. Eventually I come to realize that the reward is in traveling toward who I am. As Soren Kierkegaard said, "Now, with the help of God, I may become myself." This point underscores the need to take action.

Zone development does not come from thinking about your Zone. It comes from pursuing your Zone with actions. It comes from chasing the rainbow, from deliberately moving in the right direction. It is quite the opposite of the Greek model of cognitive priority. In the Greek view, who I am is first determined by inner, mental contemplation. "The unexamined life is not worth living," say the Greeks. "Think and grow rich" is another example of this Greek paradigm. But the Hebrew view places priority on movement, on doing something. I discover myself in the actions I take to pursue the direction my Zone points me to. I do, therefore I am.

Where do I invest the 10,000 hours? Anywhere and everywhere that moves me in the direction of who I really am. My economic value arises from my willingness to commit myself to a life governed by my Zone. The wrong normal interprets this as the question, "How do I make the most money?" It converts my Zone into a commodity for sale. The right normal views life from the perspective of the question, "How do I become useful and of value to others?" It is in my pursuit of being a benefit to others that I become most valuable and most productive. And the combination of benefit and productivity turns out to be beneficial to me too. So, if I chase the pot of gold at the end of the rainbow I will never find it. The pot of gold is not the goal. It is the beauty of the rainbow that I chase. I go after the beauty of who I am designed to be. The rest is simply commentary.

It's the bumper sticker we saw in India. "Get Rich! Die if you must!"

This radical shift in perspective has one further consequence. It views life as a verb, not a noun. Greek thinking is ultimately about accumulating things. In the Greek world, I need to possess in order to control. So, I accumulate whatever is necessary in order to reduce the risk of living. It might be money, power, prestige, influence, education or any number of other "things". To live in the Greek world is to ascribe to the motto "the one who dies with the most toys wins." It's the bumper sticker we saw in India. "Get Rich! Die if you must!"

Imagine the development of an economic engine around you

like this: When I first see a rainbow, it appears on the horizon. As long as I look in that direction, I will see it. Of course, if I turn toward another direction, I might not see the rainbow at all. So, first I must face the right way. I must set my course according to the direction of my Zone. Then I start moving toward the rainbow. At first, the horizon includes many additional vistas besides the rainbow. But as I continue toward it, I necessarily pass by those other vistas.

Things that were on the horizon recede behind me. I don't see them anymore. What I see is the constantly changing surroundings of the receding rainbow. The further I go in that direction, the more the scenery changes. Each change removes one more distraction from my view as I focus intently on only the rainbow. I am moving in a direction, but I never reach the end point of that direction. I never come to a sign that says, "You have arrived at West." Nevertheless, my travels take me through opportunity after opportunity to refine and develop my effort. And as I get better and better at traveling, I become more and more valuable to those who desire my accumulated expertise. I enjoy economic premium because I have been moving in the same direction for a long time. My journey produces fruit that nourishes others.

A Zone is a direction, not a destination.

A Zone is a direction, not a destination. As soon as someone settles down at a particular point along the journey, thinking he has arrived, his economic premium is arrested. He becomes a noun rather than a verb. He may have earned significant rewards, but his development as a person will stop.

Your Zone points the way for you to travel. It does not spell out in detail what your journey will look like. After all, it's your rainbow. Once you see the direction, apply the 10,000-hour rule and begin the journey. Take risks in the same direction, knowing the universe is conspiring with you. Become yourself – and notice how your economic premium expands.

"Go West, young man," is the watchword of Zones. Don't sit in St. Louis thinking about going West. Go! You will never discover the gold in California by sitting in St. Louis calculating what it will take to go West. And, by the way, on your journey West you will encounter all sorts of bad things. Wagon wheels will break. Rivers will have to be crossed. Mountains will have to be climbed. But each obstacle becomes a part of the journey's development, enriching your travel and honing your character. Go West! The only way you can handle all the tragedies and problems and obstacles is to know your direction and keep going. John Wooden said, "The world steps aside for a man who knows where he is going." Understanding your Zone and acting accordingly opens a pathway through the world.

> *Understanding your Zone and acting accordingly opens a pathway through the world.*

A few warnings are in order. First, successfully chasing the rainbow depends a great deal on the intensity of your desire. You must see who you can be enough so that you are willing to do what it takes to reach that horizon. Of course, the horizon will change, but intensity continues to motivate.

Second, knowing your Zone profits you nothing unless you determine to head in that direction, but taking any sort of action in the right direction will be rewarded with greater appreciation for who you are and instill greater confidence to continue. In the Bible, Moses had to return to Egypt in order to become a great leader. He could not become what he was made to be if he stayed on the backside of the wilderness. There will always be readily available excuses for staying at home.

Third, it will be tempting to "wait for it to happen." This you must resist. When there is no immediate economic gain for the engagement of your Zone, volunteer! Find a place where you can practice the essence of your life even if you aren't paid for it. You must put in the 10,000 hours and it doesn't matter if you do so as an employee or a volunteer. In fact, we might suggest that volunteering is one of the essential ingredients for Zone direction since the power of Zones is the ability to benefit another. There is no better soil for planting blessing trees than the soil of volunteer involvement.

Finally, it is possible to misunderstand the essence of a Zone while moving in the right direction. Fundamentally, operating in a Zone is benefitting someone else. The red flag question for all Zone choices is, "How will this action benefit another?" Many opportunities will present themselves as you move toward the rainbow on the horizon. Not all will be a benefit to another. True Zone development and maximum serendipity comes when you opt for the actions that have benefit for others. The wrong normal places the focus of my economic activity on me. It says that what I do must benefit me. So, I make choices based on how much money I get. The right normal first considers the impact I have on others. If it benefits them, it is a

> *True Zone development and maximum serendipity comes when you opt for the actions that have benefit for others.*

viable choice even if there is no immediate benefit to me. The wrong normal tells me to chase the rainbow in order to get the pot of gold. But Zone Phenomenon shows me that as I chase the rainbow's beautiful colors, I become a beautiful person who is of enormous value to others. It is unconscious value enhancement.

There is no protection in this world for my own well-being unless I am on the path toward the rainbow. The serendipity of my life is scattered along this path and no other. Following the Zone direction is the way to make my own luck.

Here's the reality. Some of you will read this book, recognize that it speaks to you and acknowledge that you need to find your direction and get moving – but you won't do anything about it. You will fear the consequences of change. You will worry about the bills or the family. You will succumb to personal anxiety. You won't step in the right direction. You will stay with the wrong normal because it is comfortable. But we are pleading with you not to do so.

The greatest obstacle to chasing your rainbow is you. It is your focus on yourself, your needs, your wants, your fears. With this focus, you are stuck. Knowing who you were meant to be cannot free you from this mire, because Zones are not ultimately focused on you. Zones are about your design as a benefit to others. To move in the direction of your Zone is to take your eyes off yourself and look at the horizon, a horizon filled with the needs of others for exactly what you have to offer once you are on the path. You will never be able to chase the rainbow if your eyes are constantly looking in the mirror. All that the mirror can show you is where you have been, not where you are going.

Zone Phenomenon draws a different picture of the world. It corrects the wrong normal that has become so much a part of our lives. Within this new picture, Zone Phenomenon points you in the right direction. It says, "There's the way you need to go to fit into the serendipity of the universe." But Zones framework cannot make you do anything about this. It's a compass, not a guide. You have to blaze the trail toward the rainbow, confidently anticipating that the path will get clearer and clearer as you travel. So, "What are you going to do about it?"

The greatest obstacle to chasing your rainbow is you.

We may not make you rich. But can show you how to live fulfilled.

Whoever has ears to hear, let him hear!

Chapter 17

Don't Try This At Home

OK, you get it! You understand that your Zone is the pathway, the direction, toward assured serendipity, being of optimal benefit to others and personal vocational satisfaction. You're ready to commit yourself to the 10,000 hours, including the need for volunteering. You know who you want on your personal Board of Directors[1]. You think you have some idea of the direction of your rainbow. So, why can't you do this on your own? Why can't you just take what you've read about the seven Zones and figure out which one fits you?

If you explore the Self-Help aisle of any bookstore, you won't find a book on performing open heart surgery on yourself. It's theoretically possible, but it isn't recommended. Why? Because you are more than likely to die in the process. You can't stand outside yourself to perform the surgery. It takes an outside expert to correct a heart problem. Similarly, it takes an outside expert to correct an identity problem. You are a victim of the wrong normal, and you need an expert to fix the consequences of that mistaken identity.

Fundamentally, my Zone is an action description of how I affect others, not how I affect me.

Zone Phenomenon depends on external, validated data;

data that you are not aware of because it is the data that comes from how you are perceived by another person. You can't see the picture from inside the frame. The value of my worth to another cannot be determined by me. Zone Phenomenon is not Greek metaphysics. It isn't navel gazing or meditation. It's about your effect on others and it can only be truly understood when it incorporates the actions and reactions of those outside your picture frame.

Zone Phenomenon overthrows the myth of self-help. The philosophical basis of self-help rests on the idea that I can have a clear picture of my own character and identity through a careful examination from inside. Self-help is precisely what it says. It is the self helping itself. But all self-help depends on a correct evaluation of my own condition. If I make any mistakes in self-diagnosis, the path I choose for helping myself will be flawed.

Zone Phenomenon overthrows the myth of self-help. Self help is precisely what it says. It is the self helping itself.

We have discovered that almost no one correctly identifies his or her proper Zone. Why does this happen? Because people universally project what they wish to be, not who they actually are. They do not come to the process objectively. Their evaluation of their own behaviors is always tainted by the influence of ego. That's why the first step, the Zone diagnosis, must be carried out externally.

Nearly all books and manuals that purport to help you find your true calling depend on the methodology of a cognitive retreat. The self-exams that accompany these analyses also depend on your personal evaluation of the questions. The same fatal flaws that undermine psychometrics come to bear here. I am not the best judge of my own behavior. An inward inspection of my own condition is thoroughly Greek in its outlook. It's the reason why doctors fear patients self-medicating. It's just too easy to get it wrong.

Zone Phenomenon suggests that diagnostics requires an outsider's view and that once I know the direction, I cannot plan the steps before I start taking them. In other words, I set out in a direction according to my Zone and the help of my Board of Directors[1], but it is the action itself - taking the steps, that begins to define my path. The more steps I take, the more clearly I see the path. So, instead of planning and thinking about where I need to go, I just start going and that's when I see where the path leads me. I trust in the way that I am designed – and then I do the next thing!

As a reader, your tendency will be to take this book and try to figure out on your own what you need to do next. You will apply the wrong normal to this information, converting it into a cognitive consideration of a new way to help yourself. That's what the world is teaching you. Think about it. Decide what to do. Then act.

What we are saying is that the truth is upside-down. You must first get in the game before you can understand how to play. Instead of reading about how to play tennis, watching videos about how to play tennis or mastering the Wii version of digital tennis, you must

Zone Phenomenon depends on external, validated data. You can't see the picture from inside the frame.

---

[1] Jim Collins enunciates the practical concept of each person having a personal "Board of Directors". Read more of this fascinating concept at Jim Collins website: http://www.jimcollins.com/article_topics/articles/looking-out.html

get a racquet, get on the court and start hitting the ball. You must do before you can understand, not understand before you can do. Zone Phenomenon, suggests that we are designed to be verbs, not nouns, and verbs only function as action words. A verb sitting on the sidelines contemplating life has no purpose. Furthermore, Zone Phenomenon tells us that it is ridiculous to imagine that thinking about being a benefit to others is a benefit to others. If I am going to pursue my Zone, I must act as a benefit to others. I have to get in the game!

The wrong normal considers people as nouns. Remember your eighth grade grammar. A noun is a person, place or thing. That might be true in language, but it isn't true in life. You are a verb, not a noun. You are designed to act, to do, to benefit others. It's time to throw aside the wrong normal and realize that you will only become who you are by doing who you are. Of course, verbs function best when they affect a direct object, the recipient of an action. That's why you can't do self-diagnosis. The only true measure of who you are as a verb must come from the direct objects of your actions, the outsider recipients of what you do.

You must do before you can understand, not understand before you can do.

I can't even accurately evaluate my own choices about benefitting others. I can always say, "I am heading in the direction of the rainbow," not realizing that my agendas have displaced the focus of benefitting others. That's why some people operating within their Zones can produce enormous evil. They are chasing the rainbow without consideration for the benefit to others. They are moving toward the horizon, but they have not engaged the prime directive, "How does this benefit another?"

Therefore, I need outside evaluation for two reasons: To determine my Zone and to continuously correct my direction. I can't see my own blind spots. I need a personal Board of Directors to keep me on track. That's why Zone Phenomenon is set within teamwork. Zone Phenomenon rejects the egocentric orientation of the wrong normal. I am not the center of my own universe. In fact, there is no such thing as a universe of one. I am always in a community and I need the community in order to continuously calibrate my direction. Zone Phenomenon implies continual mutual accountability.

That's why some people operating within their Zones can produce enormous evil. They are chasing the rainbow without consideration for the benefit to others.

Fundamentally, my Zone is an action description of how I affect others, not how I affect me. Moving in the direction of my Zone will increase my affect on others, push me away from an egocentric view of life and lead me toward assured serendipity. I will discover who I am in the dynamics of helping others. That's the way I was designed – as an optimally functioning contributor to community.

The next time a career planner hands you a manual about "performing open heart surgery on yourself", give it back. Go engage in serving the community – and in so doing you may enjoy finding the niche where you will be of optimal benefit to others.

EPILOGUE

NOW WHAT DO I DO?

You've reached the end of the beginning. It's the end because we have shown you what the world looks like from the new paradigm of work that delights. It's the beginning because now you have to do something about it.

There are some things you can do – and some things you can't.

We have shown that you can't diagnose yourself. You can't take what you have learned about Zones and turn it into a self-help project.

Self-diagnosis is impossible because it ignores the team aspect of the Zone and is subject to being tainted by erroneous data from the ego. Zone diagnosis must come from external validation. Self-diagnosis is dangerous to others and harmful to yourself.

Self-diagnosis leaves you in a state of misunderstood delusion You need external expertise.

The risk is too high to assume you can discover your own Zone and your Key Aptitude. If you get it wrong, you set off in the wrong direction. Not only is this a waste of time, it could passionately land you in the wrong destination. Self-diagnostics is more likely to lead you into

Zone development means assured serendipity.

error – falsely believing something about yourself. When that happens, you begin to operate according to the patterns of the wrong Zone. You damage yourself and others. You wouldn't try to do a heart transplant on yourself. Don't try to derive your own Zone either. Trust us on this one.

But once you have a proper diagnosis of your Zone and Key Aptitude, there are things you can, and must, do.

You must gather a personal Board of Directors – for calibration and accountability.

You move incrementally in the direction of the rainbow, probably including volunteer efforts as you begin to accumulate the 10,000 hours.

We put an emphasis on beginning as a volunteer because volunteering immediately involves you in the effort to measure your success in terms of being a benefit to another. It removes you from the constant temptation to measure yourself by monetary standards. The money will come. Zone development means assured serendipity. However putting the money first will cause you to focus your activity on what's good for you, and that defeats the very essence of operating in the Zone.

If you hear this message and it's starting to make sense, you'll need to find a "doctor of delightful employment".

Furthermore, when you begin to make incremental changes in the direction of your Zone, your current work may not give expression to the Zone. You may be involved in a long-term career re-direction. That's difficult to do instantaneously.

After all, you still have to pay the bills. So, we encourage you to find areas where you can volunteer in your Zone, without requiring pay. This is practicing the 10,000 hours without worrying about the immediate consequences. Just find a place where you can hone in on your design, sharpen yourself, move in the right direction, force yourself to get into the game and focus on making a difference to another. When you volunteer, you will discover that the effort allows you to gauge how effective you are in what you're doing. Of course, there are all kinds of places to volunteer, not just charitable organizations. Pick places that seem to be in the direction of the rainbow and choose to help someone else.

To operate in your Zone is to discover that work is your life's delight. That's what we want for you.

Along the way, you can put some routine feedback and calibration techniques in place, like the Birthday Plan. The Birthday Plan is simple, but effective. On your birthday, ask seven good friends or family members to give you feedback about where you were of value to them during the last year. You can also ask them to give you an honest assessment of any blind spots they may have observed in you. Techniques like this will help direct you to the one pathway that leads to your rainbow.

The Last Word

If you hear this message and it's starting to make sense, you'll need to find a "doctor of delightful employment". In other words, you'll need to get in touch with Talent Research Foundation and its team of professional practitioners. Go to www.talentresearch.org. Or email us at *livinginyourzone@talentresearch.org* and we will direct you to the nearest practitioner.

This book has introduced you to a new way of looking at work – work that revolves around "autonomy, complexity, and a connection between effort and reward" in a specific arena where what you do has intrinsic meaning – for you! To operate in your Zone is to discover that work is your life's delight. That's what we want for you.

Are you ready to get started?

Acknowledgments

This book is the culmination of more than 25 years of research into talent design - the intrinsic talent architecture innately embedded in each individual. These insights were not developed in the safe and secure world of academic research. They were developed, proven, pummeled, tried and tested in the rough and tumble of the business world. Their integrity stood and withstood the harshness of the marketplace which often is ruthless and unforgiving - "If it doesn't work, it doesn't work", even if all the academics and management guru's said otherwise.

I am grateful for the varied opportunities to work with corporate organizations and non-profit institutions - from helping CEO's pick the right individuals, in executive team building, conducting management audits, performance analysis, executive coaching and other management interventions.

I cherish the many opportunities my clients provided to work with their senior management teams - and in some cases the entire organization. One of my earliest memories of doing "talent diagnostics" was in a hardware store in 1982 in a small town, sitting on some cement bags and analyzing all the employees who worked there!

I want to express my heartfelt thanks for the many who helped to proof read the text, correct grammar and make the content more presentable. David Helmer, your eye for textual precision is astounding. Randall Croom, thanks for your expert comments and editorial inputs. Patrick Wee, I am grateful for the many days you invested in patiently plowing through the draft with Ruth and me - and bearing with my occasional impatience. Barbara, your willingness to invest your time to read and review the draft was invaluable. Mike, thanks for reading the draft aloud to ensure the flow was right - I would never have had patience for that.

Ron Harness, thanks for introducing Julian Itter - she was a God-send. Gifted, meticulous and creative - she gave the book a face it otherwise never would have had. Ruiping and Daryl, thanks for occasionally pressing the pause button on your wedding preparations to work on the typeset design. Tian Yi, grateful for your patience in re-working the textual design and your eye for detail.

Drew and Mary-Ellen, Don and Kathy your financial partnership in this book project was pivotal to make this happen. Skip and I are truly grateful. Lim Lan Chin, thank you for sharing your resources to ensure the books rolled off the press - your support was providential and timely.

Skip, my friend and brother, I can't thank God enough for you. Since our first meeting in 2005, we've connected and bonded like we have known each other since birth. You have been a precious comrade-in-arms, a trusted ally and a perceptive partner - an "ezer" (in Hebrew) I could always depend on. Your masterful command over language is exceptional and one of its kind. I've watched you in action as you re-scripted a passage and made it come alive – turning it into a literary masterpiece, for which posterity will always be grateful to you. Without you this book may never have happened!

Ruth, my wife and friend, thanks for hanging in there with me. Your sacrificial companionship means much. I may not have decked you in scarlet but I trust I have helped you discover yourself, ennobled your soul, enriched your worldview and walked with you into an "abundant life" that is more fulfilling than a billion dollars or other fanciful things our contemporaries yearn for. How on earth could I have found you? Thanks for being there.

John B. Samuel
March 2011

Made in the USA
San Bernardino, CA
29 June 2018